PRAISE FOR *THE FLY IN THE OINTMEN*

"Joe Schwarcz has done it again. In fact, he has outdone it. This book is every bit as entertaining, informative and authoritative as his previous celebrated collections, but contains enriched social fiber and ten percent more attitude per chapter. Whether he's assessing the legacy of Rachel Carson, coping with penile underachievement in alligators, or revealing the curdling secrets of cheese, Schwarcz never fails to fascinate."

—Curt Supplee, former science editor, *Washington Post*

"Wanna know how to wow 'em at a cocktail party or in a chemistry classroom? Take a stroll through the peripatetic journalistic world of the ideas and things of science brought to life by Dr. Joe. Here is narrative science at its best. The end result? In either place, scientific literacy made useful by *The Fly in the Ointment.*"

—Leonard Fine, professor of chemistry, Columbia University

PRAISE FOR *DR. JOE AND WHAT YOU DIDN'T KNOW*

"Any science writer can come up with the answers. But only Dr. Joe can turn the world's most fascinating questions into a compelling journey through the great scientific mysteries of everyday life. *Dr. Joe and What You Didn't Know* proves yet again that all great science springs from the curiosity of asking the simple question . . . and that Dr. Joe is one of the great science storytellers with both all the questions and answers."

—Paul Lewis, president and general manager, Discovery Channel

PRAISE FOR *THAT'S THE WAY THE COOKIE CRUMBLES*

"Schwarcz explains science in such a calm, compelling manner, you can't help but heed his words. How else to explain why I'm now stir-frying cabbage for dinner and seeing its cruciferous cousins—broccoli, cauliflower and brussels sprouts—in a delicious new light?"

—Cynthia David, *Toronto Star*

PRAISE FOR *RADAR, HULA HOOPS, AND PLAYFUL PIGS*

"It is hard to believe that anyone could be drawn to such a dull and smelly subject as chemistry—until, that is, one picks up Joe Schwarcz's book and is reminded that with every breath and feeling one is experiencing chemistry. Falling in love, we all know, is a matter of the right chemistry. Schwarcz gets his chemistry right, and hooks his readers."

—John C. Polanyi, Nobel Laureate

MONKEYS, MYTHS AND MOLECULES

ALSO BY DR. JOE SCHWARCZ

Is That a Fact?: Frauds, Quacks, and the Real Science of Everyday Life

The Right Chemistry: 108 Enlightening, Nutritious, Health-Conscious, and Occasionally Bizarre Inquiries into the Science of Everyday Life

Dr. Joe's Health Lab: 164 Amazing Insights into the Science of Medicine, Nutrition, and Well-Being

Dr. Joe's Brain Sparks: 179 Inspiring and Enlightening Inquiries into the Science of Everyday Life

Dr. Joe's Science, Sense & Nonsense: 61 Nourishing, Healthy, Bunk-Free Commentaries on the Chemistry that Affects Us All

Brain Fuel: 199 Mind-Expanding Inquiries into the Science of Everyday Life

An Apple a Day: The Myths, Misconceptions and Truths About the Foods We Eat

Let Them Eat Flax: 70 All-New Commentaries on the Science of Everyday Food & Life

The Fly in the Ointment: 70 Fascinating Commentaries on the Science of Everyday Life

Dr. Joe and What You Didn't Know: 177 Fascinating Questions and Answers about the Chemistry of Everyday Life

That's the Way the Cookie Crumbles: 62 All-New Commentaries on the Fascinating Chemistry of Everyday Life

The Genie in the Bottle: 64 All-New Commentaries on the Fascinating Chemistry of Everyday Life

Radar, Hula Hoops, and Playful Pigs: 67 Digestible Commentaries on the Fascinating Chemistry of Everyday Life

MONKEYS, MYTHS AND MOLECULES

SEPARATING FACT FROM FICTION IN THE SCIENCE OF EVERYDAY LIFE

DR. JOE SCHWARCZ

ECW PRESS

Published by ECW Press
665 Gerrard Street East
Toronto, Ontario, Canada M4M 1Y2
416-694-3348 / info@ecwpress.com

Cover design: David Gee
Cover images: © nicoolay/iStockPhoto; © ilbusca/iStockPhoto
Author photo: Owen Egan

LIBRARY AND ARCHIVES CANADA CATALOGING IN PUBLICATION

Schwarcz, Joe, author
Monkeys, myths, and molecules : separating fact from fiction in the science of everyday life / Dr. Joe Schwarcz.

Issued in print and electronic formats.
ISBN 978-1-77041-191-3 (pbk.)
978-1-77090-700-3 (pdf)
978-1-77090-701-0 (epub)

1. Science—Popular works. 2. Science—Miscellanea. 3. Chemistry—Popular works. 4. Chemistry—Miscellanea. I. Title.

Q162.S34 2015 500 C2014-907631-2 C2014-907632-0

The publication of *Monkeys, Myths, and Molecules* has been generously supported by the Canada Council for the Arts which last year invested $157 million to bring the arts to Canadians throughout the country, and by the Ontario Arts Council (OAC), an agency of the Government of Ontario, which last year funded 1,793 individual artists and 1,076 organizations in 232 communities across Ontario, for a total of $52.1 million. We also acknowledge the financial support of the Government of Canada through the Canada Book Fund for our publishing activities, and the contribution of the Government of Ontario through the Ontario Book Publishing Tax Credit and the Ontario Media Development Corporation.

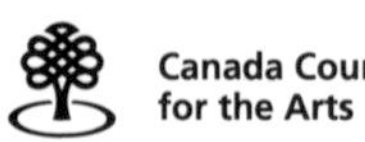

Conseil des Arts du Canada

PRINTED AND BOUND IN CANADA

Printing: Friesens 5 4 3 2 1

SWALLOW THE SCIENCE

A BACKWARD LOOK

CHEMICAL WORRIES

CHEMISTRY HERE AND THERE

HEALTH MATTERS

STRETCHING THE TRUTH

INTRODUCTION

Why "Monkeys, Myths and Molecules"? Yes, there is a link. And it is meaningful. Let me start my rationale with a question one of my students recently asked. "What is the most significant discovery in the history of chemistry?" I was really stymied because our knowledge of chemistry is based on a long string of incremental advances and it is unrealistic to label any single discovery as the most significant. There have of course been many momentous discoveries.

We can think of the smelting of metals from their ores, the development of ink, the splitting of the atom and the discoveries of the electron, radioactivity, oxygen and plastics. Dalton's theory that elements are made of atoms capable of combining in various ways to form compounds ranks up there, as does Wöhler's demonstration that chemicals made by living organisms, urea for example, can be produced from non-living substances. Haber's synthesis of ammonia led to the mass production of fertilizer that saved countless lives, as did the discovery of penicillin. But fundamental to the practice of chemistry is the understanding of the structure of molecules. So an argument can be made for August Kekulé's unraveling of the mystery of molecular structure as ranking at, or near the top, of a list of chemical discoveries. And that discovery was supposedly inspired by a dream! A dream that would lead to some monkey business.

August Kekulé's suggestion that the benzene molecule is composed of carbon atoms joined in a ring was a huge contribution to the development of chemistry. Chemists had long known that benzene consisted of six carbons and six hydrogens, but they had no idea how the atoms were arranged in space. Kekulé's solution explained benzene's behavior and provided the basis for planning other reactions that would lead to synthesizing novel dyes, drugs and plastics.

Kekulé was a dreamer, or at least he said he was. In 1890, at a meeting organized in his honor by the German Chemical Society, he recounted how two dreams had led to his being labeled the principal founder of the theory of molecular structure. The first one, in 1858, resulted in the formulation of the basic theory of how atoms can join together to form molecules. Carbon atoms were said to be tetravalent, meaning they had the ability to form four bonds, nitrogen formed three, oxygen two and hydrogen one. Most importantly, carbon atoms could link to each other, allowing for the possible formation of a huge variety of molecules. Kekulé claimed the idea came to him while riding a horse-drawn omnibus in London:

> I fell into a reverie, and lo, the atoms were gamboling before my eyes. . . . I saw how, frequently, smaller atoms united to form a pair; how a larger one embraced the two smaller ones; how still larger ones kept hold of three or even four of the smaller; whilst the whole kept whirling in a giddy dance. I saw how the larger ones formed a chain, dragging the smaller ones after them but only at the ends of

> the chains. . . . The cry of the conductor: "Clapham Road" awakened me from my dreaming; but I spent part of the night in putting on paper at least sketches of these dream forms. This was the origin of the Structural Theory.

Kekulé then went on to describe the more famous incident that supposedly occurred in 1865 in which he realized that the benzene molecule had a ring structure:

> I was sitting writing on my textbook but the work did not progress; my thoughts were elsewhere. I turned my chair to the fire and dozed. Again the atoms were gambolling before my eyes. . . . My mental eye, rendered more acute by the repeated visions of the kind, could now distinguish larger structures, of manifold confirmation; long rows, sometimes more closely fitted together; all twining and twisting in snakelike motion. But look! What was that? One of the snakes had seized hold of its own tail, and the form whirled mockingly before my eyes. As if by a flash of lightning I awoke; and this time also I spent the rest of the night in working out the rest of the hypothesis.

We will never know whether the dream stories are historically accurate or whether they represent a form of poetic license used by Kekulé to inspire scientists to think outside the proverbial box. Certainly Kekulé was not the first to whom the idea of a snake biting its tail occurred. The *ouroboros*, Greek for "tail devourer," is an ancient mystical symbol that clearly depicts a snake engaged in this activity. From the time of the ancient

Egyptians, the *ouroboros* has been used to symbolize the cyclic nature of the universe. The *ouroboros* eats its own tail to sustain its life, in an eternal cycle of renewal. Creation comes out of destruction, life out of death.

There is no written record of Kekulé mentioning dreams prior to the 1890 conference, which is a bit curious given that twenty-five years had passed since he had formulated the theory of chemical structure, supposedly based on those dreams. Even more peculiar is that four years before that conference, a drawing appeared in the *Journal of the Thirsty Chemical Society*, a publication that humorously parodied chemical discoveries, depicting monkeys in a circle grabbing each others' tails. Of course this doesn't mean that there was monkey business going on; the cartoon may have been prompted by the snake story, which perhaps Kekulé had recounted somewhere.

Speculating about the authenticity of the dream story should in no way detract from Kekulé's monumental contribution to chemistry. He surely would have received a Nobel Prize were it not for the stipulation that this prize cannot be awarded posthumously. Kekulé died in 1896, five years before the first chemistry Nobel was awarded. That prize went to Jacobus van't Hoff, who had studied under Kekulé. The second Nobel, in 1902, also went to a Kekulé student, Emil Fischer, and Adolf von Baeyer, yet another of his students, took home the prize in 1905. While Kekulé's achievements were recognized in his lifetime, he probably never dreamed how his conjured image of chains of carbon atoms closing into rings would allow his chemical descendents to dream up reactions that would change the world.

Kekulé's story about a snake grabbing its tail may well be a myth and the cartoon about monkeys forming a ring, supposedly based on the famous dream, has no basis in fact. But Kekulé's formulation of the structure of molecules has stood the test of time and has provided a foundation for separating myths and monkey business from scientific facts. And that is just what we propose to do here. Professor Kekulé concluded his 1890 talk thus: "Let us learn to dream, gentlemen, then perhaps we shall find the truth. But let us beware of publishing our dreams till they have been tested by the waking understanding." Great advice for all scientists and everyone else as well.

SWALLOW THE SCIENCE

FRUITS OF THE INTERNET

When I graduated university, my parents gave me a very special gift: a set of the renowned *Encyclopedia Britannica*! It had the answer to virtually every question I came across and was fun to just browse through. But the last time I picked up one of those heavy volumes was about twenty years ago. By then, the Internet and the tsunami of information it brought to our fingertips had appeared on the scene.

Every day I witness both the positive and negative power the web can unleash. I'm continuously flooded with questions that can be traced back to some item seen on the web. You don't even have to watch television anymore because significant clips, or entire programs, are just a few keystrokes away. A glance at just a few of the hundred or so questions that came my way during a single week affords insight into what is on the public mind. So let's have a go.

I always know what Dr. Mehmet Oz has been up to because my email inbox boils over with questions about his latest antics. Judging by the number of questions I got about monk fruit, it was clear that Oz had been trying to sweeten people's lives with this alternative to artificial sweeteners. We love sweets, but we worry, justifiably, about consuming too much sugar, and less justifiably, about its artificial replacements. The market is ripe for products that can be promoted as "natural no-calorie sweeteners." Monk fruit extracts happen to fit the bill.

Legend has it that the fruit, commonly known by its Chinese name, "*luo han guo*," was first cultivated by Buddhist monks back in the thirteenth century for its supposed fever-reducing and cough-relieving properties. Folklore also has monk fruit extending life, with claims that the counties in China where the fruit is grown for commercial purposes have an unusual number of centenarians. This has never been confirmed; neither do we know whether the fruit is routinely consumed by the population there.

Various preparations of the fruit are sold in China with claims of moistening the lungs, eliminating phlegm, stopping cough, relieving sunstroke and promoting bowel movements. While the efficacy of monk fruit as a medicine is questionable, the sweetness of its juice is not. This, however, did not arouse scientists' curiosity until the 1970s, when expanding waistlines led to expanding markets for non-caloric sweeteners. Analysis revealed that monk fruit contained five closely related compounds called mogrosides that are some 250 to 400 times sweeter than sugar. These can be extracted from the juice and processed into a powder that can be used as a sweetening agent. Because of the high degree of sweetness, very little of the monk fruit extract is required, so it can be classified as non-caloric. Although there have been no extensive safety studies, the U.S. Food and Drug Administration has classified monk fruit preparations as "generally recognized as safe," or "GRAS."

McNeil Nutritionals has now introduced monk fruit extract

as a "no-calorie sweetener" under the name Nectresse. In order to provide volume and appropriate texture, the extract is blended with a small amount of erythritol (a sugar alcohol), sugar and molasses. These contribute fewer than five calories per serving, which is the limit for a product to be labeled as "containing no calories." I'd have no problem with trying this sweetener, and I'm sure many people who worry about artificial sweeteners will pounce on it. And people are worried, sometimes for unusual reasons.

I was asked whether it is true that aspartame is made from the waste of certain bacteria. Actually, it sort of is, but that is irrelevant. Enzymes churned out by bacteria are commonly used to produce chemicals. Just think of adding *Lactobacillus bulgaricus* to milk to produce lactic acid and thereby yogurt, or using genetically modified bacteria to make insulin or human growth hormone to treat diabetes and dwarfism. In the case of aspartame, *Bacillus thermoproteolyticus* is used to join together the two amino acids that make up aspartame. I suspect the question was asked because of concern that somehow bacterial waste implies some sort of safety issue. It does not. The safety of a product does not depend on the route used to produce it; it is established by extensively studying it in the laboratory, in animal models and by monitoring its use in humans.

Then there was a question about whether ripe bananas, full of black spots, have anticancer properties. The answer is simple. No. The ripe banana silliness was based on a Japanese study reported in an obscure journal that involved injecting banana extract into the peritoneal cavity of rodents. Why this was done isn't clear. I suppose it's because nobody had done it before. Scientists will be scientists. So what did they find? A slight increase in the animals' production of tumor necrosis factor (TNF), which as the name suggests, can have antitumor effects. But it can also have negative effects, such as exacerbating arthritis. In any case, this rodent study has no relevance to humans. We eat bananas; we don't mainline them. Furthermore, the circulating email suggests that bananas contain TNF, which is nonsense. Even if they did, it would not matter because this is a protein that would be broken down during digestion. Bananas make for a great snack, but there is no point in looking for spotted ones with the hopes of preventing cancer.

Now for the really weird. A worried woman asked if it is safe to drink milk when she travels to the U.S. Why? She had heard that women who drink milk from cows treated with bovine somatotropin to increase milk production are at risk for growing mustaches. Apparently that story came from some Russian official who claimed that American milk causes women to develop male sexual characteristics. What a load of twaddle. Bovine somatotropin in not bioactive in humans. The only mustache you'll get by drinking milk is made of milk.

Next question. How do I know my answers are reasonable? Because the Internet allows me to search virtually all the published literature without leaving my desk. Clearly, the era of the printed encyclopedia is over. Although I was emotionally attached to my copy, it was taking up so much space that I decided to give it away. Nobody wanted it. A sign of the times.

ADVICE ABOUT FOOD IS SOMETIMES HALF-BAKED

Back in the early 1970s, just as I was developing an interest in the chemistry of food, I came across a witty quote by Mark Twain. "Part of the secret of success in life is to eat what you like and let the food fight it out inside." Twain was likely reacting to the plethora of health fads that were rippling through America at the time. As evidenced by a passage in his classic work *The Adventures of Tom Sawyer*, he didn't approve: "[Aunt Polly] was a subscriber for all the 'Health' periodicals and phrenological frauds; and the solemn ignorance they were inflated with was breath to her nostrils. All the 'rot' they contained about . . . what to eat, and what to drink, and how much exercise to take, and what frame of mind to keep one's self in . . . was all gospel to her, and she never observed that her health-journals of the current month customarily upset everything they had recommended the month before."

Indeed, there was health advice galore in the nineteenth century. Sylvester Graham urged people to eschew white flour, cooked vegetables and meat. Drinking water during a meal was verboten. If a vegetarian and a meat eater were shot and killed, Graham maintained, the body of the vegetable eater would take two to three times as long to become intolerably offensive from the process of putrefaction. There is no record of Graham ever putting this to a test. Dr. John Harvey Kellogg followed in Graham's footsteps, curing the rich and famous of diseases they never had with a regimen of vegetables, fruits, over-baked bread and yogurt.

Horace Fletcher, the "Apostle of Correct Nutrition," suggested that the secret to good health lay in chewing food until the last hint of flavor disappeared, and Lydia Pinkham promoted her Vegetable Compound as just the thing for "female complaints and weaknesses." Dr. James Salisbury claimed that heart disease, tumors, mental illness and tuberculosis were the result of vegetables and starchy foods producing poisonous substances in the digestive system. His solution was the "Salisbury steak," essentially fried ground beef with onion and seasonings. According to the good doctor, the steak was to be eaten three times a day with lots of water. This would cleanse the digestive system and, as a bonus, the high-meat, low-carbohydrate diet would lead to weight loss. Early shades of Atkins.

Little wonder that Mark Twain poked fun at these half-baked, contradictory fragments of advice with his suggestion to let the food fight it out once inside. That of course was pure whimsy, but foods really do duke it out, though not inside our bodies. Rather, it is in the scientific literature that dietary components vie for infamy or honor. And the biggest battles take place when the stakes are high, such as in the struggle against heart disease.

I've now been watching that battlefield for more than four decades. My bookshelves sag with dozens of books about the relationship between diet and heart disease, ranging from *The China Study*, in which Dr. T. Colin Campbell urges us to reduce blood cholesterol by eliminating all animal products to Dr. Malcolm Kendrick's *The Great Cholesterol Con* and Dr. Ernest Curtis's *The Cholesterol Delusion*, which claim that a high-fat diet does not put a person at risk for coronary artery disease and that lowering the cholesterol level with diet or drugs will not prevent heart attacks. My filing cabinets swell with the studies referenced in these books plus numerous others. One would think that a definitive conclusion about the relationship between diet and heart disease could be arrived at by digging through all this material. Alas, it is possible to find reputable studies to either support or oppose the obsession with cholesterol. When it comes to dueling studies, there rarely is a clear-cut winner.

When I began my search for light at the end of the misty tunnel of nutrition oh so many years ago, one name kept cropping up. Ancel Keys was a physiologist who had noted that well-fed American businessmen suffered a higher rate of heart disease than post-war undernourished Europeans. Keys knew that atherosclerosis was characterized by deposits of cholesterol in the walls of the arteries, and that in the early 1900s Russian scientist Nikolai Anichkov had shown a link between feeding cholesterol to rabbits and artery damage. He was also aware that in the 1940s John Gofman had identified lipoproteins as the molecules that transport cholesterol through the bloodstream and that he had demonstrated a relationship between blood levels of these lipoproteins and the risk of heart disease.

Since cholesterol is present in the human diet, mostly in fatty animal foods, Keys thought a relationship between diet and heart disease was likely. One way to explore this possibility was to compare disease patterns in countries with different amounts of fat in the diet. In his famous Seven Countries Study, Keys showed that both elevated mean blood cholesterol levels and deaths from heart disease correlated with the percent of calories attributed to fat in the diet, although there were a few exceptions. Inhabitants of the island of Crete had the lowest heart disease rate but ate lots of fat. Their fat intake, however, was mostly of the unsaturated variety found in fish and olive oil. So Keys concluded the real culprit was saturated fat and promoted a Mediterranean diet, emphasizing unsaturated over saturated fats.

Correlation, of course, is not the same as causation, and critics quickly pointed out that increased heart disease rates correlated even better with the number of radios produced or with the amount of gasoline sold. There were also questions about the reliability of death certificates to determine heart disease mortality, as well as about the calculation of fat consumption. Then there was the bothersome point of Keys choosing only seven countries when statistics about food consumption and mortality were available for at least twenty-two others. Did he leave these out because the data did not fit the straight-line relationship that was evident when only seven countries were considered? And with that salvo of criticisms, the war between the pro- and anti-fat forces was launched.

Keys correlated the risk of death from heart disease with levels of blood cholesterol and with the amount of saturated fat in the diet. This did beg for further exploration. That was undertaken with the famous Framingham study that followed more than 5,000 initially healthy inhabitants of the small Massachusetts town and confirmed that high blood cholesterol correlated, albeit weakly, with heart disease. Rarely mentioned, however, is the fact that the Framingham study found no relation between fat consumption and heart disease!

Observational studies such as Keys's and Framingham can only show associations. To prove that high cholesterol is a causative factor in heart disease, and that it is a function of diet, an intervention study is needed. A demonstration that a low-fat, low-cholesterol diet results in a drop in blood cholesterol and also parallels a decline in heart disease would constitute good evidence for recommending such a diet. In 1972, the Multiple Risk Factor Intervention Trial, cleverly abbreviated as MR. FIT, took on this challenge. Some 12,000 men at high risk for heart disease because of high cholesterol, elevated blood pressure and smoking habits were divided into two groups. One group got advice on quitting smoking, management of high blood pressure and received intensive instruction on preparing food that was low in cholesterol, low in saturated fat and high in polyunsaturated fat. The other group received no specific advice other than what would normally be offered by their family physician.

After ten years, the intervention group had reduced saturated fat intake by almost 30 percent and increased polyunsaturated fats by 33 percent, while the diet of the control group was essentially unchanged. Blood pressure was reduced significantly in the intervention group and about half the smokers gave up the habit. But in spite of the intense changes in diet, total cholesterol declined only by 7 percent. At the end of the study, there were 115 deaths ascribed to heart disease in the intervention group and 124 in the control group. Although that was significant, the result was clouded by the fact that there were 265 total deaths in the intervention group as opposed to 260 in the control group. Rigorous modification of risk factors had not provided the impactive results that had been hoped for.

Roughly at the same time as MR. FIT, the Lipid Research Clinics Coronary Primary Prevention Trial (CPPT) enlisted some 3,800 men with cholesterol levels that ranked in the top 1 percent of the population. Half were given cholestyramine, a drug expected to reduce cholesterol significantly. After ten years, the actual reduction was only 8 percent, but this did result in a reduction in non-fatal heart attacks and coronary disease deaths by 19 percent. The authors' claim that each 1 percent reduction in cholesterol would result in a 2 percent reduction in cardiac risk made for splashy headlines.

But where exactly does 19 percent come from? In the treatment group, 7 percent of the subjects died or had a heart attack, while the corresponding figure in the control group was 8.6 percent. Decrease 8.6 by 19 percent and you get 7 percent! Another way of saying this is that one would have to treat about sixty-seven people with high cholesterol aggressively to save one cardiac event. Not very impressive, but still, Keys's correlation, together with CPPT and MR. FIT, were judged to have provided enough evidence to justify making recommendations to the public about reducing blood cholesterol to reduce the risk of heart disease. That advice centered on manipulating the amount and type of fat in the diet. Incredibly, after some fifty years of research, it still is not clear just what that manipulation should be. Well, perhaps it isn't so incredible. It is very difficult to tease meaningful results out of epidemiology, especially when it comes to nutrition. Statistics about disease patterns are often ambiguous and people's recall of what and how much they ate is notoriously unreliable.

On one issue, there is agreement. Trans fats are a risk for heart disease and eliminating them amounts to good riddance. But what about saturated fats? Here the issue is not so clear. In 2010, the *American Journal of Clinical Nutrition* featured an analysis of twenty-one major studies and concluded that there was no significant evidence associating saturated fat with an increased risk of heart disease. A Japanese study in the same journal surprisingly found that as saturated fat intake increased, the risk of stroke actually decreased! On the other hand, Harvard researchers examined eight randomized controlled trials in which saturated fats were replaced with polyunsaturated fats and found a modest protection against heart disease. But even such a replacement is not so straightforward, as outlined in a 2013 paper published in the *Canadian Medical Association Journal*. Oils like corn or safflower oil, which are rich in omega-6 fats but poor in omega-3 fats, should not be promoted as reducing the risk of heart disease, while making such a claim for oils such as canola or soya, which are rich in both these fats, is reasonable.

Although the evidence for reducing the risk of heart disease by manipulating the fat content of the diet is less compelling than is generally assumed, "fat phobia" has resulted in numerous non-fat and low-fat products in the marketplace. This in spite of the fact there is no good data to show that people diagnosed with coronary disease have consumed more fat than healthy people and that more than half of all heart attack victims have normal or low blood cholesterol. Given that in low-fat foods the fat ends up being replaced by various carbohydrates, often simple sugar, we may have gone from the frying pan into the fire.

That's just what Dr. John Yudkin, a contemporary and critic of Ancel Keys, suggested. In his book *Sweet and Dangerous*, (originally published in the U.K. as *Pure, White and Deadly*) adorned with a sugar bowl sporting a skull and crossbones, Yudkin pointed out that that the correlation between sugar consumption and heart disease was stronger than that between fat and heart disease. His view was almost universally dismissed but now is being resuscitated with further evidence from Dr. Robert Lustig, who links sugar not only to heart disease but to obesity as well. Having followed the "cholesterol hypothesis" for more than four decades, I still can't come to a firm conclusion, but evidence is mounting that sugar is a greater villain than saturated fat. As Mark Twain said, "It ain't what you don't know that gets you into trouble. It's what you know for sure that just ain't so."

DINING ON LIQUID GOLD

The Greek poet Homer was in all likelihood not referring to health benefits when he called olive oil "liquid gold." The Mediterranean diet, with its health implications and emphasis on olive oil, would not hit the headlines for another 3,000 years.

Keys's findings in the Seven Countries Study were enough to convince the American Heart Association to promote a diet that aimed to reduce the risk of heart disease by reducing fat intake. To cut down the risk of heart disease, the message went, cut down on butter, lard, eggs and beef. But the fact that saturated fat intake was not the only difference between the traditional American diet and the eating habits of the Greeks and Italians was not addressed. It took the Lyons Heart Study in 1994 to demonstrate that the American Heart Association diet and a Mediterranean diet had quite different clinical outcomes.

Researchers in France investigated patients who had a heart attack and were subsequently counseled to follow a Mediterranean diet. They were encouraged to eat more fruits, vegetables and fish, less red meat, and were asked to replace butter with a special margarine that was formulated to contain the type of fats found in the Cretan diet. Why margarine? Because the researchers thought that the French palate accustomed to eating butter might accept margarine, but not olive oil. Also, the margarine contained alpha-linolenic acid, an omega-3 fat that is prominent in the Cretan diet with its emphasis on walnuts, olive oil and a vegetable called purslane. After just two years, the death rate in the intervention group was reduced by 70 percent! This approach, which controlled the type but not the amount of fat consumed, was clearly superior to the Heart Association's low-fat diet. Surprisingly, there was no difference in the blood cholesterol levels of the groups.

Then in 2003 came a prospective investigation of some 22,000 Greek adults by University of Athens researchers who developed a point system to assess how closely the subjects followed the traditional Mediterranean diet. Fruits and nuts, for example, were awarded points, and meat, poultry and high fat dairy products were not. After nearly four years, 275 of the subjects had died, most of whom had not been following the Mediterranean diet. The more closely subjects followed the diet, the less likely they were to die during the four-year trial.

While no single food was predictive of mortality, increased intake of fruits and nuts was associated with greater chance of survival, as was an increased ratio of monounsaturated to saturated fats. Such an increased ratio is a reflection of a greater intake of olive oil and a reduced consumption of meat. Unfortunately any study that relies on self-reported food intake is burdened by recall bias. In this case, the subjects were asked to report the frequency of their intake of 150 foods and beverages during the year preceding the study. Most of us would not be able to reliably recall what and how much we ate yesterday, never mind over a whole year. Furthermore it is quite possible that some of the subjects changed their diet significantly over the four-year period.

A recent randomized intervention trial published in the *New England Journal of Medicine* addressed these concerns. Spanish researchers followed 7,447 men and women who were free of heart disease but had either type 2 diabetes or at least three risk factors defined as smoking, hypertension, elevated LDL ("bad cholesterol"), low HDL ("good cholesterol"), being overweight or having a family history of coronary heart disease. Three diets were randomly assigned to the subjects: a Mediterranean diet supplemented with extra-virgin olive oil, a Mediterranean diet supplemented with mixed nuts or a control diet that just aimed to reduce fat in general.

The trial was stopped after 4.8 years, earlier than planned, because the experimental groups were showing a statistically significant reduction in cardiovascular events. Interestingly, there was very little difference in fat consumption between the groups. According to the researchers, the benefits were derived from the 50 grams (4 tablespoons) of extra virgin olive oil a day and the six servings of nuts a week. (A serving consisted of 15 grams of walnuts, 7.5 grams almonds and 7.5 grams hazelnuts.) There was also a slight increase in fish and legume intake in the experimental groups.

What's the takeaway message for North Americans? It seems that it is the type of fat rather than the amount that is important. This notion is supported by the current theory that inflammation may play a more important role in heart disease than blood cholesterol. Trans fats that lurk in many processed foods and omega-6 fats, as found in corn, soybean and sunflower oil, all popular in our diet, are inflammatory, while nut oils and olive oil have an anti-inflammatory effect.

The anti-inflammatory effect of olive oil may be double-pronged. Monounsaturated fats appear to curb inflammation but a compound called oleocanthal isolated from extra-virgin olive oil has been shown to have the same effects as the anti-inflammatory medication ibuprofen. This discovery came about in a fascinating way. Gary Beauchamp of the Monell Chemical Senses Center in Philadelphia was attending a scientific meeting in Italy, when he tried some freshly pressed extra-virgin olive oil. The experience wasn't altogether a pleasurable one, as almost immediately he began to feel a stinging sensation in his throat. As luck would have it, Beauchamp had previously worked on testing the sensory properties of ibuprofen medications and had experienced exactly the same effect. Could there be some connection between olive oil and ibuprofen? the Monell scientist wondered.

Oleocanthal was eventually isolated from olive oil and was determined to be responsible for the stinging effect. Curiously though, this compound had no chemical resemblance to ibuprofen. There seemed no doubt though that oleocanthal was the stinger, since a synthetic version added to corn oil resulted in throat irritation. Further experiments revealed that oleocanthal blocked the action of the enzymes known as COX-1 and COX-2 that are known to produce inflammation. The next question to answer was just how much oleocanthal there is in olive oil, and how much of the oil would have to be consumed for an effective dose. The answer turned out to be a great deal! A whole glass of olive oil would be needed to treat a headache. This of course is not recommended, but incorporating some extra-virgin olive oil into the diet is a good idea, especially given that the oil also contains a variety of polyphenols with antioxidant properties. This may be why some epidemiological studies have suggested that people who have a diet rich in extra-virgin olive oil may have a lower risk of breast and colon cancer.

And let's not forget that sugar, which has also been linked to inflammation, is scarce in the Mediterranean diet. So extra-virgin olive oil, nuts and low sugar and meat consumption may be the keys to cardiovascular health. But just how much protection can people with risk factors expect if they make dietary changes as in the Spanish trial? Surprisingly little. Roughly 100 people have to change their diet to prevent one cardiovascular event. Of course that is pretty significant if you are the one.

The bottom line, then, is that it pays to eat like the Cretans and other Mediterraneans. But the North American version of the Mediterranean diet will not do. Fried calimari most assuredly will not make you live longer.

SUGAR'S EFFECTS ARE NOT SO SWEET

Who would have guessed that a song by the Guess Who would become a health anthem? "No Sugar Tonight" might not have the most brilliant lyrics, but not a bad message.

"No sugar" may be impossible to achieve, but what about just six teaspoons a day? That, according to the World Health Organization (WHO), is what we should be striving for if we are to achieve the recommendation of just 5 percent of calories in our diet being attributed to sugar. We have a way to go, given that Canadians now consume a whopping twenty-six teaspoons a day! That of course is an average: teenage boys wolf down some forty-one teaspoons, while senior women only about twenty. Where is all that sugar coming from? A can of sugar-sweetened soft drink has about ten teaspoons, the same as an equivalent amount of "no sugar added" fruit juice. A smoothie can harbor more than twenty teaspoons, a serving of Froot Loops about eleven (that's 100 times more than Shredded Wheat), a candy bar around seven and a donut, four. Then there is the hidden sugar, like four teaspoons in a serving of tomato soup, and half a teaspoon in a slice of bread.

It isn't hard to see that the sugar adds up. But so what? What's wrong with sugar? After all, it's natural isn't it? And natural substances are better for us than those chemically concocted sweeteners, aren't they? Actually, no. Sugar is a problem. Of course that has nothing to do with whether it is natural or not. It has to do with what it can do as it cruises through our body. Weight gain is an obvious possibility. Extra calories translate to extra weight, and sugar can deliver a lot of extra calories. A hundred and sixty in a can of pop. You would have to run at a pace of five miles per hour for fifteen minutes to burn that off.

In everyday language, the term "sugar," normally refers to sucrose, the white crystals isolated from sugar cane or sugar beets. But to a chemist, "sugar" can mean any of a number of simple carbohydrates that have a sweet taste. Sucrose is actually composed of two sugars, glucose and fructose, joined together. Lactose, the naturally occurring sugar in milk, is made of glucose and galactose. Upon digestion, these are broken down into their components, which then enter the bloodstream. Starch, a carbohydrate composed of many glucose units linked together, is also a source of glucose upon digestion. When it comes to weight gain, the source of the sugars doesn't much matter. Carbohydrates, be they starch or simple sugars, are a problem.

Now, for the first time ever, a national regulatory agency is poised to tackle the problem. An expert committee that advises the Swedish government has recommended that new guidelines focus on a low-carbohydrate diet as the most effective method for weight loss. This is a huge turnaround given that the scientific community has largely dismissed low-carbohydrate diets as fads. However, after taking two years to scrutinize some 16,000 published studies, the Committee concluded that low-carbohydrate diets work, and that, surprisingly, in spite of being high in fat, such diets have no negative effects on blood cholesterol.

It seems that we may have been barking up the wrong tree with our calorie-counting, low-fat schemes. Diet gurus like Dr. Robert Atkins, whom we dismissed as cranks, were on the right track. It turns out that the oft-repeated dogma that weight is totally determined by calories in and calories out is theoretically sound but is of little practical significance. That's because the effective calories available from a food are not equal to the calorie content as determined by conventional experimental methods. In other words, consuming 100 calories worth of fat is not the same as 100 calories worth of carbohydrate. Fats and carbohydrates go through different metabolic pathways with different energy requirements. They also have different effects on insulin, the hormone that to a large extent determines the ratio of carbohydrates and fats the body uses for fuel. A reduced-carbohydrate diet forces the body to burn its fat stores for energy instead of glucose, the usual prime source.

But the issue isn't only about weight gain. Obesity is of course a major problem, associated with diabetes, heart disease and even cancer, but sugar seems to be a problem even aside from its link to obesity. A major study published in 2014 in the *Journal of the American Medical Association* found a clear link between added sugar intake and cardiovascular disease mortality even in the absence of obesity. Soft drinks specifically were linked to heart disease. Of course an association by itself cannot prove that sugar is the culprit, but it is suggestive, especially when one takes into account that fructose, which is released when sucrose is digested, has been implicated in causing metabolic problems.

The WHO's recommendation of 5 percent of total calories is an extreme challenge to a population now consuming about 15 percent of total calories as sugar. And it is a bitter pill for the sugar industry to swallow because such a cutback could translate to billions of dollars in lost revenue. So we will undoubtedly hear the usual arguments about moderation and how sugar can be part of a balanced diet. Well, that depends on how one determines what amounts to a balanced diet. The WHO's experts have stated that in their view a diet isn't balanced if more than 10 percent of calories come from sugar.

When making dietary recommendations, one always has to consider any potential downside. With curbing sugar intake, there isn't one. Sugar is not a dietary requirement. Of course cutting down is hard because it tastes so good. And it is also hard to know where it hides. It may be listed as barley malt, evaporated cane juice, corn sweetener, maltodextrin, brown rice syrup, molasses, dextrose, glucose and of course high-fructose corn syrup. Time to be on the lookout for all these. One easy way to cut down is to just drink water instead of pop. Life may not be quite as sweet, but it may well be longer.

Unfortunately this means limiting fruits and giving up or strictly curbing potatoes, bread, pasta, rice, cereals, ice cream, pastries and of course sugar. On the other hand you can eat all the meat, eggs, fish, cheese, nuts, yogurt, avocado and any vegetable that grows above the ground to your heart's delight. And it really means to your heart's delight. I realize that this seems paradoxical, since the usual advice is that fat intake be controlled to maintain a healthy cholesterol profile. While this seems logical, evidence now indicates that on a low-carb diet, HDL ("good cholesterol") actually increases without any adverse effect on LDL ("bad cholesterol"). Triglycerides also drop, as do blood glucose levels. The latter is important because less glucose in the blood results in less insulin being released from the pancreas, which in turn results in more fat being used for fuel. Insulin normally blocks the release of fat from fat cells. The release of fat also leads to a longer feeling of satiety, with studies showing that when people eat all they want on a low-carb diet, total calorie intake drops.

For short-term weight loss, it is clear that low-carb works best. Over the long term, all diets work about the same for the simple reason that people stray and do not adhere to them. But I suspect that when the long-term studies that are ongoing now are concluded, we will see that curbing carbohydrate intake, particularly sugar, is the most effective regimen. What about exercise? Phenomenal for overall health, but it does not have much impact on weight loss. And there is one other fly in the ointment. A low-carb diet is low in fruits, and fruits have been linked with an anticancer effect, although I suspect that eating more vegetables can compensate since these are as loaded with antioxidants as fruits, if antioxidants are indeed protective against cancer.

So what's for breakfast? An omelet will do, or yogurt with nuts and berries. Lunch and supper can feature meat, although not any of the processed varieties, with mashed cauliflower and a large salad with avocado and olive oil. As far as oils go, there is a consideration. Polyunsaturated fats have been widely promoted to lower cholesterol, which indeed they do. But there are two kinds of polyunsaturated fats — the omega-6 fats and the omega-3s. Omega-3 fats found in fish oils as well as in flax and canola oils are heart healthy, but the omega-6 oils such as corn, sunflower and safflower, while reducing total cholesterol, actually make LDL, the "bad cholesterol," more prone to oxidation. Oxidized LDL is a significant risk factor for heart disease.

For snacking in front of the TV, nuts, cheese, olives, a boiled egg or canned mackerel will do. Unfortunately, no pizza and certainly no soft drinks. Beer, tragically, as far as a low-carb diet goes, amounts to liquid bread.

I know this sounds counterintuitive, but we always say that science is a self-correcting discipline. And with the evidence we now have it is time to correct the information about recommending the low-fat diets. In any case, ever since fatphobia reared its head in the 1960s, the obesity rate in North America has skyrocketed. So clearly the low-fat option has not worked.

CONFRONTATION WITH E. COLI IS A NASTY BUSINESS

You have probably never heard of Andre Jaccard, but if you eat meat, you have likely benefited from his invention, although some would argue that the term "benefited" has to be qualified. What cannot be argued is that back in the 1970s Jaccard revolutionized an industry by patenting his meat-tenderizing machine.

Tough meat is a tough sell. And what makes for tough meat? An abundance of collagen, the robust protein that makes up what is generally referred to as "connective tissue." To make meat more tender, collagen has to be disrupted either chemically or physically. Moist cooking for a long time will do this as will aging, marinating in an acid solution, or treatment with a plant enzyme such as papain, extracted from papaya. But collagen can also be degraded by grinding, pounding or "jaccarding."

Andre Jaccard's invention was a machine that tenderizes meat by piercing it with a series of needles and razor sharp blades that surgically shred the connective tissue and thereby, at least according to the manufacturer's claim, make any cut of meat "butter tender." "Jaccarding" also allows more complete penetration of marinades and reduces shrinkage and cooking times. It is easy to see why such mechanically tenderized meat appeals to suppliers, retailers, caterers and restaurants. After all, it means being able to satisfy palates with cheaper cuts. But it may also mean exposing diners to some nasty microbes, such as the notorious *E. coli* O157:H7.

This wicked bacterium was first identified in 1982 after contaminated hamburgers caused an outbreak of severe bloody diarrhea. It hit the big time in 1993 when undercooked burgers from a fast food restaurant caused the death of four children and sickened 600 other people. That's because if the internal temperature does not reach at least 70°C (158°F), the bacteria can survive and release their toxin.

How is it that we haven't heard of this particular strain of *E. coli* until recently? Actually, that isn't so surprising given that bacteria are constantly undergoing evolution as they struggle to survive in a changing environment. This is what appears to have happened to *E. coli*, most strains of which are harmless and commonly inhabit the intestines of humans and animals. But somewhere along the line a particularly vicious strain evolved in the gut of ruminant livestock, such as cattle, deer, goats and sheep. Curiously, the bacteria do not affect the host in any detrimental way. People, however, can become severely ill should they be infected with the microbe shed by animals in their feces. That's just what we witnessed in fall 2012 when tainted beef from the XL Food plant in Alberta precipitated the largest beef recall in Canadian history.

Roughly half of all cattle shed *E. coli* O157:H7 in their feces and then end up contaminating their hides as they romp through the muck in feedlots. Then, when the hides are stripped off after slaughter, the bacteria can be transferred to the meat. Similar transfer can occur through removal of bacteria-tainted entrails. Should the contaminated meat then be ground, the bacteria can become distributed throughout.

But not everyone who became sick from meat that originated in the XL plant ate hamburgers — some apparently became ill after eating roasts or steaks. This caused suspicion to be cast on jaccarded meat, given that the process can drive bacteria from surface deep into the tissues, where they may survive, especially if the meat is consumed rare. Meat that has been tenderized in this fashion is not easy to identify, since the holes made by piercing seal up and vanish. If jaccarded cuts were labeled, as is being considered, consumers would at least be alerted to making sure that an internal temperature of 70°C is reached. Of course, there would still be no way of knowing whether meat consumed in restaurants, hotels or catered events was jaccarded.

And should you think that giving up meat offers protection from the ravages of *E. coli* O157:H7, think again. Tragically, a young girl succumbed to the effects of this bacterium, having been infected by planting a kiss on the cheek of her ailing grandfather who had been stricken after eating tainted beef in a veterans hall. The little girl lost her life to a hamburger she had never eaten! Another youngster was luckier, surviving a shutdown of her kidneys after being contaminated by *E. coli* O157:H7 at a petting zoo. Long-term consequences are, however, still possible, since the bacterium can also damage the pancreas, lungs and liver.

Staying away from contact with animals is not a guarantee against contamination either. Actually, there are more outbreaks of *E. coli* O157:H7 infection caused by produce or water than by meat. The major outbreak in 2000 in Walkerton, Ontario, that resulted in seven deaths and 2,000 people becoming sick was traced to manure from nearby farms polluting the water supply. We have seen outbreaks caused by spinach, unpasteurized apple juice and sprouts. In Europe in 2011, more than 4,000 people became sick from eating fenugreek sprouts. It seems the seeds used for sprouting had been contaminated, probably by exposure to manure. Since manure is commonly used as fertilizer, any fruit or vegetable that has been exposed, particularly if these are eaten raw, is a hazard. Obviously, it is really important to wash all produce well, paying particular attention to sprouts since these are grown under conditions well suited to the growth of bacteria.

Even cantaloupes should be washed before being cut, as the next section describes. That may sound like overkill, but it is not! Cutting with a knife can transfer bacteria from the surface to the edible portion, setting up a potential catastrophe. This isn't fear mongering. In 2011, thirty-three people died after eating cantaloupe traced to one Colorado farm that was contaminated with *Listeria monocytogenes*. Scary stuff. And we haven't even talked about fish in China that are partly fed with feces from pigs and geese, shrimp farmers in Vietnam who use ice made from bacteria-laden water or grape tomato pickers in San Juan who wipe their hands on their pants after answering nature's call in the field. Still worried about smart meters giving off dangerous radiation, artificial sweeteners or trace chemicals leaching out of water bottles?

A TALE OF TWO CANTALOUPES

This is a tale of two cantaloupes — one that killed and one that cured. Herb Stevens was a spry eighty-six-year-old who suddenly developed tremors and chills and became so weak that he was unable to get up from the toilet. And so began a downward spiral of complications that would eventually lead to his demise. Tests revealed that Mr. Stevens had been infected with *Listeria monocytogenes*, a soil bacterium commonly found in animal feces. Two weeks earlier, the retired hydrologist had eaten half a cantaloupe purchased at local Colorado supermarket, a purchase that would turn out to have lethal consequences.

Unfortunately Mr. Stevens was not the only victim; before the 2011 *Listeria* epidemic subsided, 147 people would be hospitalized and 33 would lose their lives. All had eaten cantaloupes that were eventually traced to a Colorado farm owned by brothers Eric and Ryan Jensen who were charged with introducing adulterated food into interstate commerce. The charge did not imply that they knew, or should have known, about the contamination, but as owners of the farm they were accountable for maintaining sanitary conditions. Prosecutors urged a heavy-handed approach to send a strong message to the food industry about its responsibility to reduce food-borne illness. It is indeed a critical responsibility, given that bacteria and viruses lurk everywhere in our food supply, just waiting for a chance to wreak havoc on our health. The judge in the case agreed with the prosecutors, sentencing the Jensens to five years of probation and six months of home detention. Each also was ordered to pay $150,000 in restitution and perform 100 hours of community service.

It is difficult to estimate the extent of illness caused by microbes because the vast majority of cases resolve after a brief tussle with cramps, nausea and diarrhea and never get reported. The so-called "twenty-four-hour flu" is a misnomer. Influenza is not a one-day phenomenon, but symptoms associated with food poisoning can sometimes pass in twenty-four hours. If you're lucky. If you're unlucky, contaminated food can kill. The Centers for Disease Control in Atlanta estimates that in the U.S. there are about 50 million food-related illnesses a year, with 130,000 hospitalizations and 3,000 deaths. Most of the victims are children, the elderly and people whose immune systems are compromised. Pregnant women are especially at risk, but healthy people generally are not seriously affected.

Virtually any food can be contaminated by bacteria, but cantaloupes are particularly prone because of their continued contact with the soil during growth. Furthermore, their rough skin can trap and hold bacteria, some of which can even penetrate to the inside of the melon. Just slicing a melon can transfer bacteria from the outside to the inside, which is why washing the fruit before cutting is wise. Growers are expected to minimize risk by maximizing precautions, but there are many points during processing that allow for the possibility of contamination. Investigation showed that Jensen Farms was lax in the maintenance of proper sanitary conditions.

Although the exact source of the bacteria was never pinpointed, numerous samples taken around the farm revealed the presence of *Listeria*. Contact with a hard-to-clean potato-washing machine was a possibility, as was cattle manure found on cantaloupe transport truck tires. The packing house had pools of water on the floor, and the melons were not properly cooled after coming off the fields.

Listeria is not the only bacterium that can contaminate

cantaloupes. In 2012, a Salmonella outbreak that made 261 people sick and caused three deaths was traced to Chamberlain Farms in Indiana. The victims were located in twenty-four states, a sobering reminder of how our current food distribution system can cause widespread problems. A cattle pasture next to the growing field may have been the origin of the bacteria, but once again the major problem was a bevy of inadequate practices in the packing house that ranged from lack of monitoring of wash water disinfectant levels to droppings from birds roosting in the rafters above food contact surfaces.

These outbreaks, although tragic, serve to highlight the need for vigilance in food production. At home, wash produce with running water even if it is going to be peeled. Fruits or vegetables with uneven surfaces, such as cantaloupes, can be scrubbed with a produce brush to remove microbes that are otherwise difficult to dislodge. Care should be taken not to spray water during washing because bacteria and viruses can live on surfaces for a long time. For extra safety, surroundings can be wiped with a sanitizing solution made by adding a teaspoon of bleach to a quart of water. Wipe surfaces and wait ten minutes before rinsing with clean water.

This discussion certainly is not intended to scare anyone away from eating cantaloupe — a fruit that is a good source of the antioxidants beta-carotene and vitamin C. Rather, the goal is to highlight the need for awareness about microbial contamination, the biggest concern when it comes to the safety of our food supply. One bite of contaminated food can have deadly consequences.

Obviously a cantaloupe can kill, but cure? A case can be made for one particular cantaloupe purchased in 1941 by Mary Hunt, a bacteriologist working at the Agricultural Research Lab in Peoria, Illinois. Thirteen years earlier, Alexander Fleming had discovered the antibiotic properties of a mold that had accidentally drifted into one of his bacterial cultures, and within ten years Florey and Chain had identified penicillin as the bactericidal ingredient. The search was now on to find a mold that would produce a higher yield of the substance.

Researchers throughout the world were asked to send samples of moldy fruit, grains and vegetables to Peoria for testing. Mary Hunt also took up the challenge, and on her usual shopping trips scoured produce for mold. One day she found a moldy Texas cantaloupe that aroused her interest and brought it to the lab. After cutting out the mold, she and fellow workers enjoyed the sweet taste of the historic fruit. When the mold that had contaminated the melon was steeped in a vat of corn liquor, it yielded twenty times more penicillin than any other mold tested! Within a year enough penicillin was produced to treat the vast number of battlefield infections, saving thousands of lives. "Moldy Mary," as she came to be known, had chanced upon the most celebrated cantaloupe ever grown.

DON'T EAT ARMPIT CHEESE AROUND MOSQUITOES

Have you ever wondered why mosquitoes are attracted to Limburger cheese? Probably not. It's because the cheese has the fragrance of sweaty feet! And for mosquitoes that scent is ambrosia! It's a signal that a blood meal is just one proboscis poke away.

The female mosquito needs blood to nurture her eggs, and human blood suits her just fine. But how does she locate her prey? By smell. And human feet do smell. They're inhabited by a variety of bacteria, including *Brevibacterium linens*, that metabolize the various proteins and fats in sweat, converting them to a large variety of compounds, including methyl mercaptan, hydrogen sulfide, acetic acid, isobutyric acid, valeric acid, diacetyl, acetone, acetaldehyde, phenol, indole and acetophenone. It is this combination of fragrant compounds that signals the mosquito that a meal is in the offing. And guess where else *Brevibacterium* is found? Since the nineteenth century it has been used to impart the particular flavor and the characteristic disturbing smell to Limburger cheese, named after the Duchy of Limburg where it was originally produced.

All cheese begins by curdling milk, usually with a combination of a bacterial culture and a mixture of enzymes known as rennet, originally isolated from the stomach of calves. The starter bacterial culture converts lactose, the main sugar in milk, into lactic acid, which in turn causes the milk proteins to precipitate and form curds that trap much of the milk's fat. Rennet's role is to harden the curd by forcing more proteins out of solution.

The bacteria that cause curdling also play a role in the eventual taste of the cheese. As the cheese ripens, bacterial enzymes transform the proteins and fats in the curds to a wide array of compounds. Sometimes additional bacteria are mixed in during the ripening process to impart specific flavors and characteristics. Lots of bacteria: there are some one billion *propionibacteria* that frolic in every gram of Swiss cheese, for instance, churning out propionic acid and carbon dioxide, responsible for the characteristic nutty flavor and of course the holes. Like the bacteria used to make Limburger cheese, these are also present on human skin and contribute to body odor.

So, does this mean that bacteria isolated from human skin can be used to make cheese? Not only is it possible, it has been done! Biologist Christina Agapakis, teamed up with smell expert Sissel Tolaas to make a variety of cheeses with bacteria sourced from people. They swabbed between toes, armpits, mouths and noses and then introduced the cultures into milk. Why? Not because the world is waiting for "armpit cheese." They were interested in exploring how microbes work together to create different smells and tastes and to bring attention to the important role bacteria play in our bodies and our food supply. Although many people associate bacteria with illness, they can also play a role in health. Those that may confer a health benefit on the host are known as probiotics. *Propionibacteria* are in this category. Studies have shown that they may play a role in the prevention of colon cancer by producing short-chain fatty acids that kill colorectal cancer cells.

Now to address the question of what armpit or foot cheese tastes and smells like. That depends. Different people have different microbial flora, so the taste and smell of the cheese will vary. Tasters described the armpit cheeses as having sour, floral, fruity, fresh cream or fish smells. They were far harsher in their descriptions of the foot cheeses: putrid, rotten, sweaty and fungal were the terms used. Limburger cheese odor has been described the same way. That is not to say it doesn't have its fans, but for many people the aroma is both the beginning and the end of the acquaintance. Though not in Wisconsin, the cheese capital of the U.S. There you can go into a tavern and order a Limburger on rye bread with raw onions and brown mustard. The sandwich is usually served on freezer paper and is traditionally washed down with a locally brewed beer. Some places even accompany the serving with a breath mint. Good idea.

The research into the smell and flavor characteristics of armpit and foot cheese just might earn an IgNobel Award. After all, the original research on the appeal of Limburger cheese to mosquitoes garnered the 2006 prize for Biology. The IgNobels are annual prizes given by the publishers of *Annals of Improbable Research* and are presented at a special ceremony at Harvard University. They are designed to first make people laugh and then think. The goal is to celebrate the unusual and honor the imaginative, thereby stirring people's interest in science, medicine and technology. While seemingly funny, the Limburger-mosquito connection has a very serious side. Malaria is a global scourge and any substance, be it Limburger cheese or something else, that can be used to lure the little buggers into a trap is welcome.

DR. OZ SHOULD BE RED-FACED OVER HIS PORTRAYAL OF RED PALM OIL AS A MIRACLE

Dr. Oz is a powerful guy, blessed with a name that conjures up wizardry. He just unveils his latest "miracle," which seems to happen on an almost daily basis, and people scamper off to the nearest health food store. The great Oz anointed the oil extracted from the fruit of the palm tree that grows in Indonesia and Malaysia as a wonder product that can aid weight loss and reduce the risk of Alzheimer's and heart disease. Introduced to this marvel by his guest, a homeopath, Dr. Oz excitedly gushed about its beta-carotene and "special form of vitamin E." I found this curious. Tell me, does a professor of surgery at Columbia University with more than 400 research publications under his belt really need advice on nutrition from a homeopath?

As is usually the case with Oz's miracles, there is a seed of truth that then gets fertilized with lots of verbal manure until it grows into a tree that bears fruit dripping with unsubstantiated hype. For example, one study did show a reduction in the severity of cholesterol-induced atherosclerosis in rabbits fed high doses of red palm oil. This has little relevance for humans, but magicians who pull rabbits out of hats may consider adding red palm oil to the diet of their little assistant. The red color of the oil comes from beta-carotene, the same substance that contributes to the hue of carrots and many other fruits and vegetables. It is the body's precursor for vitamin A, which makes it an important nutrient.

Unfortunately, in many areas of the developing world there is a shortage of both beta-carotene and vitamin A in the diet, leading to a high incidence of blindness, skin problems and even death. In such cases, red palm oil would be useful, but of course there are numerous other ways to introduce beta-carotene into the diet, including "golden rice" — rice that has been genetically modified to provide the nutrient. Aside from remedying a vitamin A deficiency, there is not much evidence for increased intake of beta-carotene outside of that contained in a balanced diet. There are suggestions that higher blood levels of beta-carotene reduce the risk of breast cancer in high-risk women, but the beta-carotene levels may just be a marker for a better diet.

As far as the Alzheimer's connection goes, Oz may have been referring to a study in which 74 seniors with mild dementia were compared with 158 healthy seniors. People with dementia had lower levels of beta-carotene and vitamin C in their blood. Again, this does not prove that the lower levels are responsible for the condition, they may just signal a diet that is poorer in fruits and vegetables. Tocotrienols, the "special form of vitamin E" Oz talked about, have shown some borderline effects in Alzheimer's patients at doses way higher than found in red palm oil. There is no evidence for preventing the disease.

What about the claim that red palm oil causes loss of belly fat? That seems to come from a rat study in which a tocotrienol-rich fraction extracted from palm oil caused a reduction in fat deposits in the omentum, the tissue that surrounds organs. There was no evidence of abdominal fat reduction and, furthermore, the study involved putting the animals on an unnatural and unhealthy diet. But these are not the facts that the audience was treated to on *The Dr. Oz Show*.

What the eager viewers witnessed were three visually captivating but totally irrelevant demonstrations of the purported health benefits of red palm oil. First in line was a piece of apple that had turned brown because of oxidation. This could be prevented with a squirt of lemon juice, Oz explained. Then came the claim that red palm oil protects our brain the same way that lemon juice protects the apple. This is absurd. Vitamin C inactivates polyphenol oxidase, the enzyme that allows oxygen to react with polyphenols in the apple resulting in the browning. The human brain, however, bears no resemblance to an apple, except perhaps for the brains of those who think it does. Yes, oxidation is a process that goes on in the human body all the time and has been linked with aging, but suggesting that beta-carotene's antioxidant effects protect the brain like lemon juice protects the apple is inane.

Just as zany was the next demo in which two pieces of plastic half-pipe representing arteries were shown with clumps of some white guck, supposedly deposits that lead to heart disease. Oz poured a gooey liquid, representing "bad fats," down one of the tubes, highlighting that it stuck to the goo. Then he proceeded to pour red palm oil down the other pipe and lo and behold, the deposits washed away. Totally meaningless and physiological nonsense. The homeopath then explained that saturated fats behave like thick molasses cruising through the cardiovascular system, but palm oil does not, despite being high in saturated fats. While saturated fats may lead to deposits, they do not do this by "thickening" the blood. Arterial deposits are the result of some very complex biochemistry and are not caused by "sludge" in the blood. Oz even exclaimed that this demo was indicative of how red palm oil reduces cholesterol in a month by 40 percent, better than drugs. A search of PubMed, an online compilation of medical publications, reveals no such study.

The final demonstration involved Dr. Oz lighting a candle and a flare, without wearing safety glasses mind you. The message seemed to be that the body burns most fats slowly, but it burns red palm oil with great efficiency, preventing weight gain. Where does this come from? Possibly some confusion about medium chain triglycerides, which are metabolized somewhat faster than other fats. But these are not found in palm oil. They are found in coconut oil and palm kernel oil. Oz and his homeopath expert were as confused about this as about the rest of red palm oil info they belched out.

Aside from scientists who took issue with the misleading information, animal rights groups also attacked Oz's exhortations about the benefits of the oil, claiming that it will lead to destroying larger stretches of the jungle, home to many wild creatures, including the orangutan. They maintain that when the jungle is cleared every living creature is either captured or killed and adult orangutans are often shot on sight. A tragedy. Another tragedy is that Dr. Oz could be doing so much good if he just focused on real science, as he sometimes does, instead of drooling over the latest "miracle" as presented by some pseudo-expert.

COUNT RUMFORD'S TASTEFUL INVENTIONS

I like to spice up my lectures with appropriate cartoons. When one day a student asked about my favorite, I didn't have to think too long. It's one published in 1802 by satirical cartoonist James Gillray depicting a lecture at the Royal Institution of Great Britain in which a dose of laughing gas inhaled by a volunteer makes a dramatic exit through his rear portals. An impish looking Humphry Davy can be seen preparing the gas as Benjamin Thompson, better known as Count Rumford, looks on. In 1799, Rumford was one of the founders of the Royal Institution, which aimed to make scientific study more accessible to the public. The Institution's lectures continue to this day, with the traditional Christmas presentation being the year's highlight. In 2012, for example, Dr. Peter Wothers captivated the audience with amazing demonstrations as he focused on "modern alchemy." For this ongoing effort to bring science to the public, we owe a huge debt to Count Rumford.

The title of "Count" was awarded to the American-born scientist and inventor by the Bavarian government for his help in reorganizing the country's army, establishing workhouses for the poor and introducing cultivation of the potato. Thompson had chosen the name "Rumford" after the New Hampshire town where he had been married, somewhat of a curious choice given that he had abandoned his rich wife during the American Revolutionary War, in which he sided with the British. He first moved to England and then to Bavaria, where he made a fascinating discovery in, of all things, cooking!

Rumford had a long-standing interest in heat, having experimented with gunpowder, candles, oil lamps, chimneys and industrial kilns. The Rumford chimney, which increased updraft for more efficient heating, was the toast of London. He even invented double boilers, the drip coffee pot and a machine for drying potatoes, which was basically a chamber with hot air blowing through it. Potatoes were boiled, peeled, thinly sliced and then dried. They could then be kept for years. If you're thinking potato chips, you're right.

Since time immemorial, people have cooked food in boiling water, essentially because it's easy to do. But Rumford had always wondered whether this was the ideal temperature for cooking. One day he decided to see what would happen if he placed a piece of mutton in his potato drying machine. Would the taste be different if the meat were cooked at this lower temperature? He checked on the mutton three hours later and saw that it showed no signs of being done and concluded that the heat was not sufficiently intense. He abandoned the experiment, telling the maids to cook the mutton properly. Apparently they forgot about the meat until the next morning and when they checked on it, they discovered that it had been perfectly cooked despite the fact that the fire that had supplied the heat had gone out during the night. Rumford was delighted and the concept of slow cooking at low temperature was born. The sous-vide method, currently championed by many chefs can be traced back to this fortuitous discovery made by Count Rumford.

The usual goal in cooking is to bring the food to the specific temperature at which it is perfectly "done." In traditional cooking this can be difficult to achieve because in a pan, oven or grill the outside gets very hot and may be overdone before the inside of the food cooks properly through conduction. "Sous vide," which literally means "under vacuum," circumvents this problem. Food is placed in a plastic bag, the air is removed, the bag sealed, and then immersed in hot water at the relatively low temperature of about 55°C (130°F). Removal of air from the plastic bag ensures close contact between the food and water, allowing for exact temperature control. Since the temperature is low, the cooking time is long — anywhere from hours to days. But the results, according to many chefs and foodies, are well worth it. What you get is perfectly cooked food with all the natural juices, nutrients, flavors and seasonings sealed into the food. The problem with sous vide is that low temperatures won't brown food, but this can be overcome with a quick sweep with a blowtorch or a fast sear on a grill to achieve a crust.

One concern that arises is the leaching of chemicals from the plastic bag into the food. When it comes to leaching from plastics, the chemicals most often mentioned are bisphenol A and the phthalates. Bisphenol A is used to make clear, hard polycarbonate plastics and is not used to make plastic bags. Phthalates are plasticizing agents that are added to some plastics such as polyvinyl chloride (PVC) to make them soft and pliable. PVC can be formulated into a hard plastic for such things as water pipes and records, or with the inclusion of plasticizers such as phthalates it can be made into items like shower curtains, toy duckies and food wraps. But the bags for sous vide cooking are not made of plastics that require plasticizers to achieve pliability. These bags are made of layers of polyethylene and nylon, which do not contain any plasticizers. Virtually nothing is transferred to the food from these bags. Why "virtually"? Because when any two surfaces come into contact, there will always be an exchange of some chemical. With our detection capabilities now being such that parts per quadrillion can be detected, there will always be some "contamination." Of course this does not mean that it is relevant. The presence of a chemical does not equal the presence of a risk.

The first widely produced slow-cooked food was "Rumford's soup." At the bequest of the Bavarian government, Count

Rumford looked into providing the cheapest possible nutritious food for prisoners, the poor and the army. His concoction made with pearl barley, yellow peas, potatoes, salt and beer was made palatable with long, slow cooking. I'm going to give it a try. Maybe it can even be marketed with proceeds going to the Royal Institution, which apparently is experiencing some financial problems. Wouldn't it be great if Count Rumford came to the posthumous rescue of the Institution he established? And I'm sure some cartoonist would rise to the occasion.

PEEL APPEAL

Where would you find the most famous apple tree in history? The Garden of Eden might be difficult to locate, but in any case the biblical story makes no mention of apples. The fruit of "the tree of knowledge of good and evil," forbidden to Adam and Eve, is not named.

Curiously, the Latin word "malum" means both "evil" and "apple," which may have led to the association of the renowned tree with apples by early Christians. John Milton then cemented the image in his classic "Areopagitica" with the lines "It was from out the rind of one apple tasted that the knowledge of good and evil, as two twins cleaving together, leaped forth into the world."

While the Garden of Eden is mythical, Woolsthorpe Manor in Lincolnshire, England, is as real as can be. And in the garden of that manor stands the apple tree that can indeed be deemed the most famous in history, for it was a falling apple from that tree that led Isaac Newton to eventually formulate his classic theory of gravity. No, the apple didn't fall on his head as many cartoons depict, but the falling apple story is true. Newton himself gave the account to a number of people including William Stukeley who wrote the first biography of Newton in 1752.

According to Stukeley, Newton described the event to him in his own words: "Why should that apple always descend perpendicularly to the ground? . . . Why should it not go sideways or upwards? but constantly to the earth's center? Assuredly the reason is that the Earth draws it." And so began the formulation of the theory of gravity as finally stated in Newton's classic work, *Principia*: "Every particle of matter in the universe attracts every other particle with a force that is directly proportional to the product of the masses of the particles and inversely proportional to the square of the distance between them."

But had it not been for the plague, Newton's mind may never have been put into motion by the falling apple. In 1665, the terrible disease struck England, forcing many institutions, including Cambridge University, where Newton was studying, to be closed. He returned home to Woolsthorpe and whiled away the time sitting in the garden and made the chance observation that would lead him to conclude that the same physical laws that governed the falling apple also applied to the movement of the heavenly bodies. What history does not record is whether Newton ate the fallen apple. Maybe he did. And maybe he ate the fruit regularly. After all, he lived to the age of eighty-five, stunningly long for the times!

Apples have long been associated with good health, dating back to such early medical notables as Hippocrates and Galen, who suggested eating apples after a meal to aid digestion. The world's first medical school, the Schola Medica Salernitata established in the ninth century, taught that cooked apples were useful for disturbances of the bowel, lungs and nervous system. Master surgeon John Gerarde in 1597 recommended apples for the treatment of "hot stomacke." And then in 1886, a Welsh magazine offered up the proverb "eat an apple on going to bed and you'll keep the doctor from earning his bread." This was eventually shortened to the popular, oft-repeated phrase "an apple a day keeps the doctor away."

That of course is wishful thinking — no single food has such miraculous properties. But we are learning more and more about the benefits of apples, particularly of eating them with the peel on! Apple peel is particularly rich in polyphenols, compounds with significant antioxidant activity. Antioxidants are reputed to have all sorts of health benefits because they neutralize the potentially damaging reactive oxygen species that are byproducts of our body's use of oxygen. While specific antioxidants in pill form have been disappointing, there is a wealth of evidence indicating that consumption of fruits and vegetables is protective against disease. Perhaps other modes of action of plant chemicals are more important than antioxidant activity. Some studies suggest that polyphenols may alter gene expression, others indicate that they may have "prebiotic" activity, meaning that they can modify the bacterial flora in our gut, which in turn can have health benefits.

There is obviously a great deal of interest in further exploration of the link between naturally occurring chemicals in plants and our well-being. One approach is to study the effect of exposing cells cultured in the laboratory to specific food extracts, with cancer cells being of particular interest. Recent studies have shown that apple peel extracts can significantly reduce the proliferation of a variety of cancer cells, including those isolated from breast, prostate and liver tumors. Certainly this is interesting and worthy of further research, but such experiments offer no evidence for the use of apple peel extract in humans for the prevention or treatment of cancer. On the other hand, we do have plenty of epidemiological studies that have linked apple consumption with a reduced risk of lung cancer, cardiovascular disease, chronic obstructive pulmonary disease and stroke.

Understandably, the antiproliferative effect of apple peel extract on cancer cells has boosted research into its potential use as a dietary supplement. One particular extract of organic apple peel, marketed as AppleBoost, has already been shown to increase antioxidant activity in the blood of people who consumed the powder, which is easily blended into smoothies or yogurt, over a twelve-week period. What we now need are studies to see if the antioxidant activity translates into health benefits over the long term. In the meantime, the best advice is to wash apples well and eat them with the peel. The more the better, except for people with inflammatory bowel disease, whose condition may be exacerbated.

Finally, it seems that it is time to discard the current practice of discarding the massive amount of apple peel generated by the apple juice, apple sauce and baked apple product industries. Perhaps producers should take a bite out of the fruit of the tree of current knowledge. Unlike in the Garden of Eden, ignorance is not bliss.

CHEW ON THIS

I'll tell you up front that I don't like chewing gum. I'm familiar with the studies that have shown chewing may reduce tooth decay, help with weight management and even reduce stress, but I'm not won over. And it isn't because I'm worried about the "carcinogens, petroleum derivatives, embalming fluid ingredients or chemicals that cause diarrhea or mess with our digestive system." I don't buy these type of accusations that permeate the Internet, authored by scientific luminaries with self-conferred titles such as "The Food Babe."

My aversion to gum probably traces back to elementary school when one of my teachers had a unique punishment for anyone caught chewing gum in class. The criminal had to climb on a chair and recite: "The gum-chewing student and the cud-chewing cow differ somehow. I know, it must be the intelligent look on the face of the cow!" Ever since witnessing such a "sentence" being carried out, I can't look at a masticator without comparison to a cow, decidedly to the animal's advantage. Memory sure is a mysterious thing. And therein lies a gummy story to chew on.

Back in 2002 researchers at Northumbria University in England assigned seventy-five subjects aged twenty-four to twenty-six to either chew gum, mimic chewing without gum, or not chew at all while performing both short- and long-term memory tests. Gum chewers scored significantly higher. Although the robustness of this study has been criticized, it did unleash speculation about why chewing gum may aid memory. For one, research has shown that chewing gum increases blood flow to the brain and activates the frontal and temporal cortex, probably by enhanced oxygen transport. Since these regions are known to play a role in cognitive function, increased memory seems a possibility. Another option is context-dependent memory, implying that information is more easily recalled in an environment similar to the one experienced while learning, particularly if a smell is involved. For example, students studying while exposed to the scent of chocolate perform better when exposed to the same scent as they write exams.

In 2011 Dr. Matthew Davidson of Stanford University's School of Medicine explored the memory enhancement effect further, adding another twist. He fortified the gum with substances that have been associated with improved cognition. Many studies have suggested improved alertness with caffeine, so it was a natural additive. In fact the U.S. army has introduced caffeinated gum in military rations. Davidson also added an extract of the *Ginkgo biloba* tree along with vinpocetine derived from the lesser periwinkle, both of which have been shown to enhance blood flow to the brain. Also included was an extract of a creeping herb known as *Bacopa monnieri*, which in at least one placebo-controlled, double-blind study was shown to improve learning rate and memory. Rosemary and peppermint were also added, mostly to take advantage of their strong scent that may enhance recall. The sixty-two participants were divided into three groups; they chewed either the "Think Gum," ordinary bubble gum or nothing while engaged in learning as well as during recall.

The results were not exactly spectacular but the herbal-caffeinated gum chewers did perform better on most tests, in some cases by as much as 30 percent, than the non-chewers. Bubble gum chewers did only marginally better than non-chewers, implying that the effects noted were due to the gum's additives rather than to chewing. While Think Gum may be a helpful study aid, the product's advertising slogan of "stop cheating, start chewing" is a bit hard to swallow.

What is even harder to swallow is the diatribe directed at

chewing gum on some websites and circulating emails. Here is one gem: "Gum is typically the most toxic product in supermarkets and is likely to kill any pet that eats it . . . loaded with harmful ingredients and chemicals." (Hmm, I wonder what sort of ingredients don't contain chemicals?) One of the "harmful ingredients" highlighted is BHT (butylated hydroxytoluene), a preservative. The basis of the scare? It is found in embalming fluid! Yes it is, but so what? BHT is an approved food additive that prevents fatty substances from reacting with oxygen. It is and is actually sold in pill form in health food stores as a dietary antioxidant.

Another "toxin" we're supposed to worry about is vinyl acetate. This is indeed a worrisome substance if exposure is significant, but that isn't the case with gum. These days, gum base is mostly made of synthetic rubber rather than natural substances like the sap of the sapodilla tree, commonly known as chicle. The synthetics include styrene-butadiene rubber, polyethylene and polyvinyl acetate (PVA). In theory, PVA may contain traces of vinyl acetate, the chemical from which it is made, but the amount is trivial. The devil, as they say, is in the details.

The scares about gum that go around today, however, can take a back seat to the one that rocked the Egyptian city of Mansura in 1996, when loudspeakers blared out warnings to young girls about the evils of chewing gum "laced with aphrodisiacs capable of transporting the most innocent female into a sexual frenzy." Sold under the brand names Aroma and Splay, the gum was said to be an Israeli attempt to corrupt Egyptian youth. Laboratory analysis by Egypt's Ministry of Health found nothing other than the usual ingredients, but that didn't stop authorities in Mansura from closing any kiosk that dared sell the gum that supposedly caused young ladies to engage in immodest activities.

Talking about immodest activities, how about a gum that claims to increase bust size? Bust Up gum contains miroestrol and deoxymiroestrol, two estrogen mimics found in the underground tubers of a plant called *Pueraria mirifica*. Lack of evidence for the reputed benefit notwithstanding, it is curious that some women who may be concerned about trace amounts of estrogen mimics leaching out of plastics will swallow significant amounts of phytoestrogens when it comes to trying to improve their appearance.

And now for a final bit of nonsense. Calcium casein peptone-calcium phosphate (CCP) is an ingredient in Trident gum. Various alarmists warn about it with inspired comments like "casein is a milk-extractive that was linked with the Chinese baby formula poisonings." Casein is indeed a protein found in milk, but it has nothing to do with the poisonings that were due to formula being adulterated with melamine in an attempt to increase protein content. CCP actually remineralizes tooth enamel by delivering calcium and phosphate beneath the tooth's surface. The scare about it is just another example of the rampantly galloping chemophobia we're witnessing. That stresses me. Maybe I should chew gum.

AN UPLIFT FROM PISTACHIOS

Remember when it wasn't hard to determine if someone had been into the pistachio bowl? They'd be caught red-handed! That's because until artificially colored foods became a pariah, pistachio nuts, which are actually not nuts but the seeds of a fruit, often used to be colored red. Exactly why that was the case is a matter of some controversy.

Some suggest that when pistachios were first imported into North America back in the 1930s, mostly from Iran, the shells tended to be blemished as a result of hand-picking. Since Americans didn't care for blemished food, the pistachios were dyed red. Others suggest that the red color was added to distinguish the newly introduced nuts from other varieties to attract attention. Another possibility is that in Iran the nuts were traditionally soaked in brine and then roasted in the sun, which resulted in a pinkish-colored shell and thus importers added red dye to achieve a uniform product. The fact is that nobody really knows how the tradition started, or indeed what dye was used, although some accounts make reference to a vegetable dye, probably beet juice. With concerns being raised about food additives, the red pistachios have mostly disappeared, although a few companies still produce them for nostalgic consumers. The vast majority of pistachios sold in North America now come from California, and instead of attracting consumers with color, producers hope to attract them with science. The hook is a possible benefit in the prevention of heart disease, and believe it or not, help with erectile dysfunction.

Nuts are low in saturated fats, high in monounsaturates and rich in antioxidants, so it comes as no great surprise that epidemiological studies have demonstrated a link between increased nut consumption and reduced risk of cardiovascular disease. Pistachios have a chemical profile similar to nuts and have therefore been studied in terms of reducing cardiovascular risk. In one small study, subjects were asked to consume 40 grams, 80 grams or no pistachios daily. The pistachio consumers lowered their LDL cholesterol (the "bad" type), but interestingly, there was no difference between the 40- or 80-gram consumers. So one pistachio snack seems to be enough — more is not better.

But does this extra consumption not lead to weight gain? Apparently not. A study in China examined the pistachio effect in some ninety subjects diagnosed with metabolic syndrome, which is a combination of disorders that increase the risk of developing cardiovascular disease and diabetes. Although there are some variations in the definition of metabolic syndrome, it basically means a large waist circumference combined with any two of elevated triglycerides, reduced HDL cholesterol (the "good guy"), raised blood pressure, raised fasting glucose or previously diagnosed type 2 diabetes. In the Chinese study, subjects consumed no pistachios, 42 grams or 70 grams for twelve weeks. There were no changes in body mass index or waist to hip ratio. Curiously, there was also a slight improvement in triglyceride levels in the 42-gram group but not the others.

Pistachios have also been the subject of a study by Dr. James Painter of Eastern Illinois University, who coined the term "Pistachio Principle," referring to an effect by which the body is fooled into eating less by using visual cues. Painter found that when subjects were offered either shelled or in-shell pistachios, there was a significant difference in calorie intake but no difference in degree of satisfaction. Subjects offered shelled pistachios consumed 41 percent fewer calories, the assumption being that the reduction in calories was due to the additional time needed to shell the nuts or perhaps to concern about the extra volume perceived when the nuts were in their shell. In a subsequent study, subjects were allowed to consume in-shell pistachios at their leisure with the shells being discarded in a separate container. For half the subjects, the container was periodically emptied. In the case where the containers were not emptied, subjects consumed fewer nuts without any difference in fullness or satisfaction ratings. The conclusion was that leaving pistachio shells as a visual cue may help consumers consume fewer calories.

The point of the Pistachio Principle is not to encourage eating fewer nuts. Rather, the point is that altering environmental cues can lead to satisfaction with less food. For example, studies show that large package size increases caloric consumption by some 22 percent. Buying single-serving potato chips and small-size candy bars as opposed to family-size bags reduces consumption. Of course it is far better to forget the chips and candy bars and eat pistachios instead.

Now we come to the most intriguing effect of pistachio consumption, an improvement in erectile dysfunction. Pistachios are rich in arginine, an amino acid that leads to nitric oxide production, a chemical known to improve blood flow to the penis. A paper published in the *International Journal of Impotence Research* contends that pistachios may raise more than just hopes for men suffering from ED. Subjects who consumed 100 grams of pistachios a day for three weeks showed an improvement based not only on subjects' reports, but on measurements of penile blood flow as well. Perhaps this is why historically pistachios have been considered to be the food of royalty.

According to legend, the Queen of Sheba pronounced that pistachios were to be an exclusive food offered only to the royal household. When the Queen heard of the great wisdom of King Solomon, she journeyed to see him with gifts of spices, precious stones and gold. As the Bible tells us, King Solomon reciprocated and "gave to the queen of Sheba all her desire, whatsoever she asked." Not sure what she asked for, but at least according to the 1959 epic film *Solomon and Sheba* starring Yul Brynner and Gina Lollobrigida, the couple's meeting blossomed into romance. Perhaps pistachios had been part of the Queen's gifts.

A passion for pistachios was also exemplified by Nebuchadnezzar, the ancient king of Babylon. It is said that in his hanging gardens he had planted pistachio trees. Akbar the Great, a Mogul Emperor, would hold royal feasts that were fit for a king. He usually served chickens that had been fed pistachio nuts for at least six to eight weeks to enhance their flavor. And maybe enhance something else as well. Some marketer is probably thinking right now of coloring pistachios a viagric blue.

THE MANY FACES OF CASTOR OIL

I think the first medicine I ever heard of was Ricinus, a liquidy concoction with which my mother plied me when she suspected I was constipated. I can't imagine why as a child I would have had such a problem, since our diet in Hungary back then included generous doses of goose fat that should have allowed everything to slide through at a pretty regular pace. Why though am I telling you about my youthful bowel habits? Because these memories were triggered by a question that has come up about a substance called polyglycerol polyricinoleate that appears on the label of some chocolate bars. As one might guess from the name, there is a connection to Ricinus, forged through the castor bean plant, botanically known as *Ricinus communis*. It is a pretty plant, sometimes grown ornamentally, but mostly cultivated for the seeds in its fruit that can be pressed to produce an oil for use in paints, glues, brake fluids and various lubricants. The same oil can also be used to produce polyglycerol polyricinoleate, an emulsifier that is now used extensively by chocolate manufacturers.

Chocolate lovers of course look for taste, but they also seek smoothness. A gritty product just won't do. And therein lies a challenge. Chocolate is basically a mixture of cocoa butter, cacao particles, sugar and, in the case of milk chocolate, milk. The texture of the final product depends on how well these components can be blended together, which in turn depends on how effectively the friction between the ingredients can be reduced. This is where emulsifiers come in. These chemicals serve as internal lubricants, leading to a smoother consistency and easier flow when the chocolate is melted. That is of great importance when producing chocolate coatings.

The classic emulsifier used in chocolate manufacture has been lecithin, mostly derived from soy oil. But it is increasingly being replaced by polyglycerol polyricinoleate because of its greater effectiveness at reducing the viscosity of the chocolate and, perhaps more importantly, because it allows for less cocoa butter to be used and therefore makes for lower-fat chocolate. Increasing cocoa butter content increases smoothness, but cocoa butter is expensive. The use of polyglycerol polyricinoleate allows for the production of cheaper chocolate without sacrificing texture. Of course, consumers wonder if anything else is being sacrificed, such as safety. A virtually unpronounceable chemical name raises skepticism in many minds, especially when people discover that castor beans contain one of the most toxic natural substances known, a protein called ricin.

This chemical is so toxic that prior to World War I the U.S. investigated its use as a coating for bullets. When ricin gets into the bloodstream, it can kill in incredibly tiny doses. Since powdered ricin can be inhaled, it was also investigated during World War II for possible use in cluster bombs. The Soviet KGB put ricin to a practical use, supplying the Bulgarian secret police with tiny ricin pellets, which were used to assassinate dissident Georgi Markov in London in 1978 with a modified umbrella that used compressed gas to fire the pellet. And today there is concern that terrorist groups are looking to extract ricin from castor seeds. But there is no need to worry about chocolate. Ricin is not soluble in fat at all and does not end up in the oil when the castor beans are pressed. And of course the polyglycerol polyricinoleate produced from the oil has undergone the stringent regulatory process required for a food additive. Pronouncing the complex term may be a challenge, but polyglycerol polyricinoleate can be consumed safely.

For me, castor oil conjures up memories of bitter taste. For some Italians, it might be a reminder of some bitter history. The expression "*usare l'olio di ricino,*" meaning "to coerce or abuse" traces back to Benito Mussolini's Fascist Blackshirts, who forced political dissidents to drink large quantities of castor oil as a means of torture and humiliation. Since the victims were usually beaten as well, it was said that Mussolini's power was backed by "the bludgeon and castor oil."

It is unfortunate that castor oil's name has been tainted in this fashion because the oil has a variety of important uses. The Blackshirts abused it, but in appropriate amounts, the oil can used to treat constipation and can also serve as a moisturizing agent for the skin. Industrially, castor oil can be reacted with compounds called isocyanates to yield polyurethanes, widely used plastics. Automobile manufacture uses large quantities of polyurethanes which traditionally have been made from petroleum derivatives. Castor oil is a renewable resource, so it provides for a "greener" process.

The oil can also be converted into lithium 12-hydroxystearate, a high-performance lubricant grease, as well as into sebacic acid, used in the production of a type of nylon. The oil itself is an effective lubricant and was widely used in the rotary engines of Allied airplanes in World War I and is still used to lubricate model airplane engines as well as race car engines. The Castrol Company that produces various lubricants derives its name from castor oil, which in turn gets its name from the Latin "*castor*" for beaver. That's because of its use as a replacement for castoreum, an extract of the perineal glands of the beaver. Beavers use castoreum as a territorial marker, but cosmetic chemists have found that adding it to perfumes maintains the fragrance for a longer time. Castor oil can act similarly but is cheaper to produce.

While the oil will lengthen the time before a fragrance evaporates, it won't lengthen eyelashes despite the claims of some "holistic" practitioners, one of whom told me that beavers have shiny, long fur, so castor oil will do the same for eyelashes. This is where our funny bone gets tickled. "*Castor*" is beaver in French. This ill-informed practitioner thinks that castor oil comes from beavers, perhaps conjuring up some mental image of the animals being squeezed to extract their oil.

THE COLOR PURPLE

"I wouldn't drink that juice if you paid me!" And so began my conversation with a fellow traveler sitting next to me at the airport as we waited for our flight. She had glanced at my laptop and saw an item I was perusing with the headline "Purple tomato juice from genetically modified fruit engineered for health benefits." Courtesy required some sort of response, which I made with trepidation, having a feeling about the direction the discussion was about take. I thought about mentioning that I wasn't in the business of paying strangers to try unusual beverages, but I finally I came up with the more benign "And why not?" As I suspected, that unleashed a torrent of rhetoric about evil multinational companies foisting untested genetically modified foods on us. "What health benefits?" she answered. "Those foods are making us sick."

I wasn't really surprised by the onslaught, after all, this is not a unique view. But, no, genetically modified foods are not making us sick. The studies attesting to their safety are overwhelming. But can we say with certainty that no untoward effects will ever arise? Of course not. That is a naïve expectation that science can never meet. All we can do is come to some conclusion about risks and benefits based on the current state of knowledge.

My new acquaintance was mistaken about genetically modified foods making us sick, but she was correct in questioning their health benefits. Up to now, it is only farmers who have benefited from crops resistant to herbicides and insects grown from genetically modified seeds. Indeed, one of the reasons for public skepticism about genetic modification is the lack of any obvious benefit to the consumer. The purple tomato is a step, admittedly a small one, toward demonstrating that health benefits are possible, and that pursuing the technology along these lines is worthwhile.

The purple color is due to an accumulation of anthocyanins, compounds that occur widely in nature, although not in conventional tomatoes. They are responsible for the stunning hues of autumn leaves and the various colors of flowers, fruits and berries. But plants do not produce anthocyanins as entertainment for our eyes — it is reproduction they have in mind. Colorful flowers and fruits attract insects and animals to spread pollen and seeds. The tomato, however, produces very little of these compounds, relying mostly on lycopene from the carotenoid family to attract attention. But use a clever bit of genetic engineering to introduce a couple of genes from the ornamental snapdragon and, presto, you have a purple tomato, and an obvious question. Why would anyone want such a thing? It all comes down to paying attention to the growing body of evidence that the optimal diet is mostly plant based.

Numerous studies have linked a lower rate of heart disease, cancer, obesity and diabetes with a largely plant-based diet. Is that because plants contain special disease-preventive compounds or because a meaty diet contains disease promoters? Or is it a combination of these factors? Based on laboratory studies, animal experiments and human epidemiological data, the anthocyanins have stirred interest as potential disease-preventing compounds. In cell cultures they can be shown to scavenge free radicals, stimulate the production of detoxicating enzymes, reduce the proliferation of cancer cells and interfere with the formation of blood vessels that supply nutrients to tumors. Animals with chemically induced cancers fare better when given an anthocyanin-rich diet. While human epidemiological studies have not shown a cancer-preventative effect for anthocyanins, they have for heart disease.

The Nurses Health Study monitored the medical status of more than 93,000 nurses who periodically filled out food frequency questionnaires over eighteen years. Based upon the foods consumed, researchers calculated the amount of anthocyanins in their diet and found that nurses with a high intake had a reduced risk of having a heart attack. Most of the anthocyanin intake was accounted for by berries, with the data showing a roughly 30 percent decrease in risk for those who consumed more than three portions a week compared to those who ate the berries less than once a month. A number of other chemicals found in plants were also assessed, but only anthocyanin intake was associated with a reduction of heart attack risk. Of course plants contain thousands of compounds, and it is impossible to disentangle their effects. Anthocyanins may just be markers for the presence of other active compounds. A further complication is that more than 600 anthocyanins are known to be present in plants, possibly with different health effects.

But if anthocyanins as a class do indeed have a protective effect, the pertinent question is whether the amounts found in the genetically modified tomato are significant. It turns out that they may well be. The samples with the highest content weighed in at 450 mg of anthocyanins per tomato. That's almost as much as a cup of blueberries (500 mg) and far more than strawberries. But if you really want to load up on these chemicals, go for black raspberries at 850 mg per serving, or seek out chokeberries or elderberries at a whopping 2,000 mg per cup. Alternatively, you'll be able to drink a glass of purple tomato juice, if and when it becomes commercially available. It is noteworthy that in the Nurses Study, health benefits were noted with as little as 35 mg of anthocyanins consumed every day.

The new-fangled tomatoes were developed in England by Professor Cathie Martin at the John Innes Centre in Norwich, but because of an anti-GMO atmosphere in Europe they are being grown experimentally by New Energy Farms in Ontario. Juice from the tomatoes will be sent to Professor Martin for testing of health effects with seeds removed to ensure that there are no genetically modified components in the finished product. Such effects have already been noted in animals. Mice reared on a diet of genetically modified purple tomatoes lived 30 percent longer than mice consuming red tomatoes. Tests also show that the purple tomatoes have anti-inflammatory effects.

As research zeroes in on the specific phytochemicals responsible for the healthy nature of a mostly plant-based diet, we are likely see further developments in genetic engineering geared toward improving the nutritional content of our food supply.

I tried to explain all this to my seatmate, who still maintained that she would steer clear of any genetically modified foods since she had "no need of the genes they put into purple tomatoes." Having made her point, she proceeded to take a sip from the soft drink she had been nursing and ripped open a bag of potato chips. Not an anthocyanin in sight. Sigh.

A CULTURE LESSON

Genghis Khan's armies supposedly lived on it, Nobel laureate Élie Metchnikoff thought it was responsible for the long life of Bulgarian peasants and Dr. John Harvey Kellogg used a special enema machine to pump it into his patients' bowels. Today, shelves in the dairy aisle are brimming with dozens of varieties promising not only great taste but good health. It's yogurt! Labels on packaging clamor about their contents being low-fat or sugar-free or antioxidant-rich or crammed with live cultures or all of the above. And then there is the current darling of consumers, Greek yogurt.

Basically yogurt is milk soured with lactic acid produced by the action of certain bacterial enzymes on lactose, the sugar naturally present in milk. The acid changes the structure of casein, the protein in milk, causing it to form an insoluble curd that is then suspended in a liquidy portion called whey. The bacterial action also breaks down any remaining lactose into its components, namely glucose and galactose, meaning that people suffering from lactose intolerance will have a lot less trouble with yogurt than with milk. Lactose intolerance is characterized by the lack of lactase, an enzyme in the digestive tract that breaks down lactose into its absorbable components. If the enzyme is absent, lactose travels to the colon, where it can trigger diarrhea. The colon also harbors bacteria capable of digesting lactose, producing gas in the process. The result is abdominal discomfort.

In all likelihood, yogurt was an accidental discovery made when naturally occurring bacteria invaded milk, possibly from goatskin used to make bags for carrying milk. Pliny the Elder in the first century AD already described that "barbarous nations" knew how "to thicken milk into a substance with an agreeable acidity." These barbarians had undoubtedly learned that a fresh yogurt could be made by adding a bit of yogurt from a previous batch. Indeed, this is just the way yogurt is made today. A bacterial culture is added to milk that has been heated to kill off any undesirable bacteria. Greek yogurt differs from regular yogurt in being thicker due to removal of much of the whey. Traditionally this has been accomplished by straining through cheesecloth, although today some processors use a centrifuge. Another route to "Greek style" yogurt involves adding milk protein concentrate and thickeners such as pectin or inulin to regular yogurt.

Removal of the whey has some nutritional benefits. Since the product is more concentrated, it has more protein, about 15 grams compared with 9 grams for the conventional variety. That can help with weight control because protein is filling. Since the straining process removes a lot of the soluble carbohydrates, Greek yogurt has only about half the carbohydrates found in regular yogurt. Some calcium is lost in the whey, but Greek yogurt is still a good source of the mineral with a serving supplying about 20 percent of the recommended daily intake. Keep in mind that if made from whole milk, any yogurt will be high in saturated fats, so it is best to stick to the versions made from low-fat or skim milk. And of course to ones that are made without added sugar.

Now for the downside of Greek yogurt. While it may be more friendly to our health, it is not so friendly to the environment. There are two basic issues. One, Greek yogurt generates large amounts of acidic whey that has to be disposed of, making for an environmental challenge. The whey can't be dumped into water systems because when it biodegrades it uses up a lot of the water's dissolved oxygen leading to the destruction of aquatic life. Some of the whey can be blended into feed and fertilizer or potentially be used to provide protein for infant formula and body-building supplement manufacturers but more whey is produced than can be economically used. The second problem is that it takes about four times as much milk to make Greek yogurt as an equivalent amount of regular yogurt. That's a problem because milk production itself takes a toll on the environment, due to the production of large amounts of greenhouse gases. Whenever the required fertilizers, pesticides and feed are manufactured and transported, carbon dioxide is released. Producing cows is an energy-intensive process in the first place, consuming about 50 percent more energy than pork or poultry production. And there is also a lot of energy used when dairy animals are eventually slaughtered for meat as well as when their calves are converted to meat.

Dairy farms also use a great deal of electricity because of milking machines, cooling the milk and the heating the water needed to wash the equipment. Processing, transport and packaging the milk also requires energy. Then there is the emission of methane by the cows, a more potent greenhouse gas than carbon dioxide. On top of it all, there is the issue of nitrous oxide, another greenhouse gas, being released from ammonium nitrate fertilizer. Some people argue that organic milk is the way to go because of energy savings due to eliminating fertilizers and pesticides. Also on organic farms cows graze on clover-based pastures, which is beneficial since clover can take up nitrogen from the air and convert it into compounds in the soil that can be used as fertilizer. Furthermore organic herds forage more than conventional cows, suggesting that less feed such as soy, which requires an energy input for growing, needs to be imported for the production of organic milk.

Now what about the nutritional difference between organic and conventional milk? Last December many newspapers featured headlines along the lines of "Fresh Research Finds

Organic Milk Packs in Omega-3s," sending consumers scurrying to the organic aisle in the supermarket. Omega-3 fats have developed an aura of health in spite of murky evidence, but it is true that milk from cows feasting on grass and clover contains more omega-3 fats than milk from grain-fed animals.

However, the omega-3 fat here is alpha-linolenic acid (ALA), which is not the one that has been linked with health benefits from eating fish. That, though, is hardly the point. Press reports hailed the finding of "60 percent more omega-3 fats in organic milk." Whoaaa! Talk about a misuse of numbers. The organic milk had 32 mg of omega-3 fats per 100 grams of milk versus 20 for conventional. Indeed, a 60 percent difference, but totally inconsequential! When we talk about the benefits of omega-3 fats, if there actually are any, we are talking about needing hundreds, if not thousands of mgs per day. As mentioned, there are benefits to organic milk, but touting it as a source of omega-3 fats is based on a nonsensical milking of the data.

“NATURAL” . . . OR IS IT?

Put the term “natural” on a food label and cash registers start to ring. About that there’s no controversy. But there’s plenty of controversy about what “natural” means, for the simple reason that there is no accepted definition. The most common perception is that natural substances are those that exist without any human intervention. Rocks would clearly be natural. Ditto poison ivy. But what about a tomato grown without any fertilizers or pesticides? That tomato is the product of years of cross breeding and would not exist without human intervention. So is it “natural?” The complexity of this issue is highlighted by the controversy that surrounds one of the most popular flavoring agents in the world, vanillin.

Europeans first learned about vanilla from Spanish Conquistador Hernan Cortez who had been taken by a beverage served to him, supposedly in a golden cup, by the Aztec Emperor Montezuma. “*Chocolatl*” was made from cocoa beans, spices and the cured, ground pods of the vanilla plant. Christopher Columbus was also a fan, declaring *chocolatl* to be a “divine drink which builds up resistance and fights fatigue.” He claimed, “A cup of this precious drink permits a man to walk for a whole day without food.”

Vanilla is still a mainstay in many chocolate products, but the flavoring can also be found in a variety of baked goods, colas and of course ice cream. And it can be found in the middle of a debate. Just when is vanilla natural and when is it not?

An extract of the vanilla bean contains more than 200 compounds with one, vanillin, contributing most of the flavor. Vanillin has antimicrobial properties and the plant probably evolved to produce it as a protection against bacterial invaders. In the vanilla bean, vanillin is tied up, bound to a sugar molecule, which explains why a freshly picked bean has no smell. After the beans are picked, they are immersed in hot water to initiate various enzymatic reactions, including one to disengage vanillin from the glucose to which it is bound. The beans are then "sweated" in a wooden box for a couple of days, followed by first drying in the sun and then on racks. Finally storage for three months in a closed box develops the full flavor and aroma characterized mostly by vanillin. "Natural" vanilla extract is then made by soaking the beans in alcohol to transfer the flavor and aroma compounds to the solution. Obviously "natural vanilla extract" involves a lot of processing.

Although vanilla is grown in the Caribbean, Tahiti, Indonesia, Central America and Madagascar, there isn't nearly enough to meet the needs of the world. And it is not only palates that are pleased by vanilla. Noses too. Many perfumes have vanillin in their base note. Some brains, penises and urinary tracts also clamor for vanillin. Well, not exactly for vanillin, but for drugs that are made from it. L-dopa to treat Parkinson's disease, papaverine for erectile dysfunction and trimethoprim for urinary tract infections, all of which can be synthesized from vanillin.

Cultivating the vanilla plant is very labor intensive because the flowers have to be hand-pollinated within twelve hours of opening since the bees that are the plant's natural pollinators are not found outside Central America, the vanilla orchid's original home. Chemists have, however, come to the fore and developed several ingenious methods for synthesizing vanillin. Eugenol found in cloves, nutmeg or cinnamon can be converted to vanillin as can guaiacol that is isolated from pine tar. The most economical commercial method produces vanillin from phenol, a compound sourced from petroleum. Lignin, from wood processing, is also a viable starting material for vanillin synthesis. Japanese researcher Mayu Yamamoto has even produced vanillin from lignin isolated from cow dung. Vanilla produced by this method is unlikely to be embraced by consumers given that the raw material is sort of off-putting, and since chemical reactions are involved, the final vanillin would have to appear on labels as "artificial flavor."

Let's get one thing straight. The properties of vanillin do not depend on its origin. There is no difference between vanillin extracted from the vanilla bean or that made from lignin in cow dung or synthesized from pulp and paper waste. At least there is no chemical difference. But there is a difference in perception. Consumers these days are looking for "natural" ingredients, and seeing the term "artificial" on a label hinders sales. But flavors extracted from vanilla pods can cost up to $4,000 a kilo, more than a hundred times the cost of synthetic vanillin, so it isn't surprising that natural vanilla flavor extracted from the bean supplies less than 1 percent of the total vanillin demand. Since greater cultivation of the vanilla orchid is not viable, the industry is looking for alternate ways to produce vanillin that can be labeled as natural.

Yeasts and bacteria are natural living factories that convert nutrients into a variety of metabolites. Several bacteria, such as special strains of *Amycolatopsis* or *Delftia acidovorans* can crank out vanillin when nourished with ferulic acid, a compound abundantly found in the cell walls of plants. But they don't do it very efficiently, partly because they also use vanillin as a source of energy as soon as it is formed. *E. coli* bacteria, on the other hand, do not produce vanillin and therefore have no vanillin degradation pathway.

This is where synthetic biology enters the picture. Genes that code for the enzymes responsible for vanillin production in *Amycolatopsis* or *Delftia* can be isolated and inserted into *E. coli*, which then will produce vanillin in commercially viable amounts. Similarly, common baker's yeast can be genetically engineered to produce vanillin. The industry argues that since bacteria and yeasts are "natural," the vanillin they produce should also appear on food labels as "natural flavoring." However, some consumer organizations contend that since genetic engineering is involved, such vanillin is not natural. There are some yeasts, though, extracted from rice bran, that do produce vanillin from ferulic acid without any need for genetic manipulation. Such vanillin can be marketed as "derived by a natural process," but of course there is still plenty of technology involved.

While there is no safety issue here, the taste of pure vanillin is not the same as that of natural vanilla extract since the latter contains a host of compounds other than vanillin that contribute to the flavor. And then there is another troublesome point: cheap vanillin made possible through synthetic biology can interfere with the livelihood of vanilla farmers for whom money does grow on trees.

A BACKWARD LOOK

BIBLICAL STORIES CAN IGNITE SCIENCE

Moses "looked, and behold, the bush burned with fire, and the bush was not consumed." That passage from Exodus is one of the most famous ones in the entire Bible! After all, it was from that burning bush on Mount Horeb that God spoke to Moses, telling him that he had been chosen to lead his people out of slavery in Egypt.

Searching for possible scientific explanations for Biblical phenomena is an interesting pastime. Of course that is all it is, because for those who have faith that Biblical accounts are based on true miracles, no scientific explanation is necessary. And for those who are skeptical that the Bible is factual, no scientific rationalization is needed for events they believe never occurred. Whatever one's point of view, Biblical stories can serve as a springboard for leaping into some captivating science.

Suggestions have been made that the *Dictamnus albus* plant, found throughout northern Africa, is a candidate for the burning bush. In the summer, the plant, also known as the "gas plant," exudes a variety of volatile oils that can catch fire readily and may give the impression that the bush is burning. So was Moses witnessing the combustion of a mix of terpenes, flavonoids, coumarins and phenylpropanoids? An interesting hypothesis about the burning bush, but one that can be readily doused.

The plant's volatile oils do not catch fire spontaneously; they need a source of ignition. Moses is unlikely to have been walking around with flint stones looking for bushes to ignite. And when the vapors coming off the *Dictamnus albus* plant do ignite, the flash lasts just a few seconds. Had the flash managed to set the leaves on fire, the bush would certainly have been consumed. So if Moses really did see a burning bush that was not consumed, well, maybe he was seeing things. At least that is the opinion of Benny Shanon, professor of cognitive psychology at the Hebrew University of Jerusalem.

Professor Shanon suggests that Moses may have been having a hallucinatory experience. And he bases that theory on his own fling with plants that can alter consciousness. It seems Shanon was once invited to a religious ceremony performed by natives of the Amazon. He had the opportunity to taste a potion made from the ayahuasca plant, and off he went on a hallucinogenic trip that he described as having spiritual connotations! It isn't clear exactly what he meant by that, but it's clear he liked the experience because he claims to have repeated it hundreds of times, even writing a book on the subject. If it happened to him, it could have happened to Moses, he suggests, perhaps somewhat tongue in cheek.

The problem is that ayahuasca is a tropical vine found in the jungles of the Amazon, not in the sands of the desert. However, there is plant that grows in the Sinai and the Negev with similar properties. And that is *Peganum harmala*, also known as wild rue. Like ayahuasca, the plant's seed capsules contain a number of alkaloids, such as harmine, vasicine and harmaline, that can affect the mind. These compounds interfere with the activity of an enzyme known as monoamine oxidase (MAO), which is involved in the breakdown of dopamine, serotonin, phenylethylamine, tyramine and melatonin — all compounds that play important roles in our nervous system. These monoamines, as they are called, increase in concentration in the presence of monoamine oxidase inhibitors (MAOI), such as the ones found in the seedpods of the wild rue. The result can be a consciousness-altering experience. In fact, monoamine oxidase inhibitors are used as medications to treat depression by boosting levels of dopamine and serotonin, which are involved in mood regulation. Whether the Israelites used psychoactive plants in religious ceremonies is debatable, but apparently modern-day Bedouins who wander through the same desert where the Biblical accounts place Moses do partake of wild rue.

Although rue does contain psychoactive substances, its effects are very mild compared with ayahuasca. That's because in the Amazon the natives have learned to combine ayahuasca extracts with those of another plant known as chacruna. This is a rich source of dimethyltryptamine (DMT), a potential hallucinogen. DMT is a monoamine, normally broken down by monoamine oxidase, but when chacruna is combined with the monoamine oxidase inhibitors in ayahuasca, a powerful hallucinogenic effect is produced. Moses, however, would have had no access to chacruna.

Still, Professor Shanon has another ace up his sleeve. He suggests that the acacia tree also has psychedelic properties. And that tree is frequently mentioned in the Bible and was supposedly the type of wood used to make the Ark of the Covenant as well as Noah's famous ark. Interestingly, the ancient Egyptian goddess Iusaaset was associated with the acacia tree, which the Egyptians referred to as the "Tree of Life." This may have been because all species of acacia are known to harbor DMT, either in their bark, roots, leaves or fruit. Acacia trees were also used by the ancient Egyptians to make a hallucinogenic wine called yrp, which when consumed during sacred ceremonies caused the appearance of Iusaaset, "the great one who comes forth." Once more Professor Shanon claims personal experience, having consumed an acacia extract that produced effects similar to ayahuasca.

So was Moses just having visual and auditory experiences when he heard the word of God emerge from a burning bush? Sounds a little far-fetched. But there is no doubt that entheogens, which are defined as psychoactive substances used in various religious and shamanic ceremonies, have a long history. Psilocybin and *Amanita muscaria* mushrooms, cannabis, peyote cactus, ergot fungus and of course ayahuasca have all been used by various cultures in a spiritual context.

Whatever the truth may be about Moses's adventures, the biblical account has allowed us to delve into some fascinating facets of the botanical world, including the ingenious way some natives learned to produce a hallucinogenic experience by combining plants that harbor natural monoamine oxidase inhibitors with those that feature natural monoamines. Obviously, sometimes faith can trigger science. And indeed, sometimes the wonders of the world as revealed by science can trigger faith.

A CHERISHED POSSESSION

When I open my office door in the morning my eyes invariably dart to one of my prized possessions, a historic Kipp's apparatus that is as closely associated with chemistry as the iconic Bunsen burner. Sometimes referred to as a Kipp generator, it's a device that was invented around the middle of the nineteenth century by Dutch pharmacist Petrus Jacobus Kipp for the production of small volumes of gases. Constructed of three glass bulbs stacked one on top of the other, it sort of resembles a snowman.

The glass bulbs are connected in such a way that introducing a solid chemical into the middle one and an acid into the top bulb allows for a controlled production of a gas that can be drawn off through a valve. To produce hydrogen sulfide, for example, ferrous sulfide, familiar to many people as "fool's gold," is combined with dilute hydrochloric acid.

But why would anyone want to produce hydrogen sulfide, famous as the classic odor of rotten eggs? Because this gas can be used to detect the presence of certain metal ions in solution. Upon reaction with hydrogen sulfide, these ions will form different colored precipitates. A reddish precipitate indicates the presence of mercury, a black precipitate may signal the presence of lead, an orange one is indicative of antimony, and a bright yellow precipitate is evidence of arsenic sulfide.

Today various instrumental methods are available, but back in the nineteenth century precipitation with hydrogen sulfide was the prime technique for detecting toxic metals that were sometimes used to adulterate food. Lead oxide, for example, was used to give Gloucester cheese a red hue, lead chromate intensified the yellow color of mustard, pickles were commonly greened up with copper sulfate and copper arsenite was used to make candies more appealing. The presence of these adulterants could be detected with hydrogen sulfide produced by a Kipp generator.

Until the late nineteenth century, adulteration was common. Chalk or potassium aluminum sulfate were added to flour to whiten bread and plaster of Paris, sawdust or mashed potatoes were used to increase the weight of a loaf. Used tea leaves were boiled with sheep dung and ferrous sulfate then colored with Prussian blue (ferric ferrocyanide) or verdigris (copper acetate) before being resold. Coffee grounds were rejuvenated with gravel, chicory or the dried root of wild endives. Even more troublesome was the use of strychnine or an extract of the South Asian fishberry instead of hops to impart a distinctive bitterness to beer. The fishberry was so-called because it contains picrotoxin, a poison used to stun fish. Red wine was adulterated with bilberry or elderberry juice. Coloring sauces for potted meats with iron oxide was common. Today the food industry is carefully regulated but nefarious activities cannot be totally precluded. There have been cases of paprika being adulterated with red lead and chili powder with Sudan 1, a potentially carcinogenic dye used in shoe polish.

The public was first made aware of the dangers that lurked in food in 1820 when chemist Friedrich Accum published his classic work *A Treatise on Adulterations of Food and Culinary Poisons*, with a cover duly adorned with a skull and crossbones. Accum applied his chemical knowledge to rooting out the criminals who adulterated food. He knew, for example, that a blue color forms when starch is combined with iodine and used this reaction to show that cream was often thickened with wheat starch or rice powder. Accum also identified copper by the deep blue color formed when it reacted with an ammonium hydroxide solution, and recognized the presence of lead by virtue of its formation of a black precipitate with hydrogen sulfide. This was dangerous work because it preceded the invention of Kipp's apparatus, which produces hydrogen sulfide at a controlled rate. Up to that time the gas was made by combining iron sulfide and hydrochloric acid in a flask equipped with a tube through which the gas would emerge, sometimes in a dangerous fashion. It doesn't take much hydrogen sulfide to cause a calamity.

At a very low concentration of 50 parts per million (ppm) in air, hydrogen sulfide irritates the eyes, nose and throat. At 100 ppm it becomes dangerous to life. A further problem is that at that concentration, the gas disrupts the olfactory sense so the smell disappears and levels can rise to a potentially lethal concentration of 250 ppm without the exposed person being aware. Hydrogen sulfide is a component of natural gas and is also produced by the bacterial decomposition of manure, so workers in the oil and gas industries as well as farmers who have unfortunate encounters with manure pits can be exposed to dangerous levels.

Of course, most people try to avoid exposure to hydrogen sulfide. But not everyone. Japan has been struck by a wave of hydrogen sulfide suicides. The methodology is no big secret, as anyone who has studied a bit of chemistry, or has read mystery books, knows. Just mix some sort of sulfide with an acid, and presto, hydrogen sulfide bubbles out. For some strange reason, calcium sulfide is sold in Japan as "bath salts" and mixing this with an acidic toilet bowl cleaner has become a preferred method of suicide. Thinking themselves to be honorable, some Japanese suiciders have taken to posting "beware of hydrogen sulfide" signs on their doors after several instances of neighbors and would-be rescuers being overcome by the gas.

Such an episode occurred in Ontario recently, when two police officers were overcome by fumes when they tried to stop a man from killing himself. He had posted a warning sign on the window but nevertheless the officers risked their own lives in an attempt to stop him. Sadly they were unsuccessful and ended up needing hospital treatment. The exact method of generating the gas was not revealed for fear of giving ideas to copy-cat suiciders, but the police subsequently did issue a warning that "members of the public who come across a 'rotten eggs' smell when finding someone who appears to be sleeping in a vehicle or other enclosed space are advised to immediately contact police." No need to be scared of my Kipp generator though. It sits empty on a shelf, a silent reminder of an era when hydrogen sulfide was used not to kill people but to protect them from food adulteration.

EDISON'S FOLLY

Some memories remain firmly etched in your mind. Like walking into Yankee Stadium for the first time and seeing that wide expanse of green grass, the famous façade, the monuments in center field. I was already a big Yankee fan in 1965 when I visited the "house that Ruth built" for the first time. Actually, Yankee Stadium could just as well have been labeled the "house that Edison built," given that in 1922 it was Edison's Portland Cement Company that supplied the concrete used to build the stadium. That famous landmark also turned out to be the last major construction project for a company that was born out of what has been called "Edison's Folly."

Seeing the term "folly" associated with the name of the man who invented the phonograph, made the first practical lightbulb, built the first central power station and had more than 2,300 patents to his name probably comes as a surprise. But Edison's venture into the cement business turned out to be a resounding financial failure. It all began in 1881, when the celebrated inventor's curiosity was aroused by a large bank of black sand while on a fishing trip to Long Island. He packed some into a bait bucket and took it back to his lab, where it ended up under his desk.

One day Edison accidentally dropped a magnet into the bucket, and when he retrieved it, noted that black grains stuck to the magnet while sand dropped away. The grains turned out to be magnetite, a form of iron oxide. Coincidentally, just a year earlier, Edison had patented a technique for magnetic ore separation, a result of his continuing quest for the ideal lightbulb filament. He now he jumped at the chance to put his technology into practice. The "wizard of Menlo Park" was already concerned that the scarcity of iron ore in the Eastern U.S. was causing an increase in the price of steel needed for his electric generators. He purchased the Long Island beachfront and began preliminary experiments to extract the iron from the black sand. Unfortunately the iron content of the sand turned out to be too low and the project had to be abandoned, prompting a trade journal to label Edison's attempt to extract iron from its ore with magnets as "Edison's Folly." The inventor took this as an insult and determined to show the world that the metal could indeed be extracted using his technology, maybe not from black sand on beaches, but from low-grade iron ore.

Edison focused on an abundant supply of ore around Ogdensburg, New Jersey. Here he built what at the time was the largest ore-crushing plant in the world. The ore was pulverized with giant rollers and the resulting dust dropped down huge towers, where the iron was captured with electromagnets. Right from the start, the operation was plagued by equipment breakdowns, dust that clogged the machinery and a low yield of iron. Edison, however, persisted, improved the machinery and poured more and more money into his Ore-Milling Company, predicting that he was going to do "something so different and so much bigger than anything I've ever done before that people will forget that my name was ever connected with anything electrical."

But that was not to be. Even though Edison managed to bring down the price of his iron, he couldn't compete with the product produced from the high-grade ore that was discovered in the great Mesabi Iron Range in Minnesota. Finally, in 1899 the Edison Ore-Milling Company closed for good. Edison had lost a fortune but apparently was not too bothered. "It's all gone, but we had hell of a good time spending it," he crowed.

The inventor, though, was no quitter. His company had been selling the leftover sand from the milling process to cement manufacturers, who were fond of the fine particle size. Why not forget about the iron and capitalize on the sand, Edison thought? So he opened the Edison Portland Cement company to supply the material that was much needed by the burgeoning concrete industry.

Cement is a complex mixture of compounds dominated by calcium silicates that form when limestone, sand and clay (a blend of various metal silicates) are heated in a kiln to a high temperature. The final product has the form of marble-sized pieces and is known as "clinker." This is cooled, ground and mixed with a small amount of gypsum (calcium sulfate) to produce Portland cement, named after its similarity in color to limestone quarried from the English Isle of Portland. The main use of cement is in the production of concrete, which is made by combining cement with water and various aggregates such as pebbles, gravel or shale. At this point, the concrete can be poured, and then through a series of very complex reactions between water and the cement's calcium silicates, a hard, stone-like material forms. Concrete is indispensable to modern life, being the material from which buildings, bridges, roads, sidewalks, pipes, dams and skyscrapers are constructed.

Edison recognized the potential of concrete and planned on revolutionizing the industry with his "poured cement house." He created a set of two iron molds, one inside the other, so that a whole house could be continuously poured, complete with walls, floors, stairways, bathtubs and even picture frames for about $1,200. He also envisioned concrete furniture, refrigerators, pianos and phonograph cabinets. A few of these houses were built and are still in use, but the whole venture did not pan out because the molds had to be bolted together from some 2,300 pieces and proved to be too expensive. Edison's cement company finally fell victim to the Great Depression and went into bankruptcy just a few years after his greatest project, the construction of Yankee Stadium. When the stadium was renovated in 1973, the walls made of Edison's concrete had held up so well that they did not have to be touched thanks to the innovations he had made in producing concrete.

Even Edison's failed ore-crushing venture would eventually yield benefits. It yielded forty-seven patents, many of which set forth ideas still in use today. *Iron Age* magazine, which had blasted "Edison's Folly," eventually editorialized that Edison's only mistake at Ogdensburg had merely been being twenty-five years ahead of his time.

NAPALM HORRORS

"I love the smell of napalm in the morning . . . that gasoline smell . . . smelled like victory." A classic line from *Apocalypse Now*, the iconic film about the horrors of the Vietnam War. And napalm is a true horror. Who can remain untouched by the famous Associated Press photo of a young naked Vietnamese girl's panic-stricken expression after being splashed by the fiery contents of a napalm bomb? Nine-year-old Kim Phuc miraculously survived but required extensive treatment and seventeen surgeries. She became a living symbol of the lunacy that is war and now, as a Canadian citizen, works tirelessly to help children victimized by hostilities.

Using fire in warfare not a novel idea. As early as the seventh century, Byzantine soldiers used primitive pumps to douse the enemy with burning liquid. The exact composition of this "Greek fire" has been lost to history, but it likely was some sort of combination of saltpeter, sulfur, oil, quicklime and pine resin. The resin made the concoction sticky, increasing the chance of setting fire to the target. Greek fire was really a forerunner to napalm.

During World War II, a number of scientists were approached by the National Defense Research Committee in the U.S. to help with the war effort. One of these was Professor Louis Fieser of Harvard, who had already made a name for himself by developing a laboratory synthesis of vitamin K. He was also the first to determine that cancer could be induced by specific chemicals and in 1929 managed to synthesize dibenzanthracene, the first known pure carcinogen. Fieser had also worked with lapachol, a compound isolated from the lapcho tree, which was thought to have some potential as a cancer treatment. Lapachol also demonstrated anti-parasite properties, so it was a candidate for malaria treatment, sorely needed by troops fighting in malarial regions.

Quinine, the traditional drug used to combat malaria had to be extracted from the bark of cinchona tree, but cinchona plantations had fallen into Japanese hands. Because Fieser had some experience with anti-parasitic compounds, he was asked to mount a crash program at Harvard to find a replacement for quinine. Originally there were hopes for lapinone, a compound synthesized by Fieser based on the lapachol model, but it did not prove to be effective enough in humans. Fieser had greater success with the other project he was asked to tackle, the improvement of flame throwers. Such weapons had been developed during the World War I but were not very effective because the gasoline they spewed stayed liquid and dripped off targets. The army was looking for some way to make the fuel thicker.

Some attempts at jelling gasoline with rubber latex had been made, but the Japanese had overrun many rubber plantations, so Fieser took another approach. He found that mixing a combination of the aluminum salts of naphthenic and palmitic acids with gasoline did the job. "Napalm," the name coined from the first letters of the critical ingredients, naphthenic acid from petroleum and palmitic acid from palm oil, turned out to be a brutally effective weapon, allowing flame throwers to jettison a blazing goo for impressive distances. Napalm was also readily incorporated into bombs to set cities afire, with the added military advantage of asphyxiating the enemy. As the napalm burned, it used up the oxygen in the air and produced large amounts of toxic carbon monoxide.

During the war, much of Dresden and Tokyo were destroyed by napalm bombs. These were large bombs dropped from airplanes. Not little bombs carried by bats. Yes, bats! Believe it or not, using bats as flying incendiary bombs was not only thought of, but actually tried. And Louis Fieser was involved in this curious project as well. The original idea came from Lytle S. Adams, a Pennsylvania dental surgeon who was outraged by the Japanese attack on Pearl Harbor. Some sort of retaliation had to be mounted! Adams had previously witnessed the famous nightly emergence of thousands of bats from Carlsbad Caverns in New Mexico and this gave him an idea. He approached Harvard professor Donald Griffin, a recognized expert on bats who had achieved fame by elucidating how the creatures use reflected sound waves to pinpoint objects in their flight path. Adams suggested equipping thousands of bats with small time-delay incendiary explosives and releasing them at night from airplanes over Japanese cities. The bats would seek shelter for the day and roost in the many wood-frame buildings, hopefully igniting thousands of fires.

Strangely enough, Griffin did not dismiss the idea, saying that "this proposal seems bizarre and visionary at first glance but extensive experience with experimental biology convinces the writer that if executed competently it would have every chance of success." President Roosevelt got wind of the scheme and agreed: "This man is not a nut. It sounds like a perfectly wild idea but is worth looking into."

Louis Fieser did more than look into it. He actually developed miniature incendiary devices that could be attached to bats and even organized field tests. One of these was disastrous, as an airfield's buildings and even a general's car were ignited by the flaming bats. Although specialized cages to transport the bat bombs on airplanes were actually manufactured, the plan to unleash a bat attack against Japan was abandoned when the development of the atom bomb became a research priority.

Napalm, however, was not abandoned. Indeed, after the war the American military continued research into trying to enhance napalm's effectiveness. This resulted in the substitution of polystyrene and benzene for the salts of naphthenic and palmitic acids. The novel mix, which burned much longer, was called Napalm B, although it lacked the components of its namesake. This became the incendiary weapon of choice during the Vietnam War, inflicting hideous burns, often on innocent civilians. Fieser, who developed lung cancer because of his smoking and became an anti-smoking crusader, never expressed regret for developing napalm. "I have no right to judge the morality of napalm just because I invented it," he maintained. But I have no problem in judging. Napalm doesn't smell like victory. It reeks of decomposing morality.

FLYING HIGH WITH GOLDBEATER'S SKIN

To put it bluntly, Louis XVIII stank. It wasn't from improper hygiene, although French kings weren't particularly noted for their love of baths. The culprit was the gangrene oozing from his legs! When poor Louis finally passed away, the odor was so foul that something had to be done before the body could be properly prepared for burial. So, who you gonna call? Antoine Germain Labarraque, that's who. The man who had solved the problem of stench in the *boyauderies*, the factories where animal guts were turned into strings for musical instruments, sutures and a useful substance called "goldbeater's skin." Labarraque knew just what to do. He covered Louis's body with sheets soaked in a sodium hypochlorite solution (bleach) and made the undertaker's job bearable.

The processing of animal guts into string is older than recorded history. The Ebers Papyrus, compiled around 1550 BC, describes Egyptian surgeons using dried intestine for suturing wounds. And the gold leaf that adorns ancient sarcophagi was made with goldbeater's skin, the incredibly thin but strong outer membrane of calf intestine. Gold is the most malleable of all metals, meaning that it can be beaten into extremely thin sheets. Egyptian craftsmen managed to produce gold leaf that was an almost unimaginable 0.000125 mm thick! To achieve this, layers of gold sheets were piled on top of each other and then laboriously beaten. Separating the layers with "goldbeater's skin" was critical to prevent the leaves from fusing together. But preparing the skin was not easy. It involved separating the outer membrane of the intestine and treating it with a mild alkaline solution to allow residual fat to be scraped off. The Egyptians probably used potash (potassium hydroxide), made by adding calcium oxide (lime) to potassium carbonate. Calcium oxide forms when calcium carbonate (limestone) is heated, while potassium carbonate is found in wood ashes.

The exact methods of producing gold leaf and goldbeater's skin were kept secret by goldsmiths, who usually passed the skills down through their families. But the stench associated with processing animal intestines could not be kept a secret. That's why around 1820 a French association dedicated to encouraging industrial development offered a financial reward for a method that would reduce the disturbing odors that plagued the industry.

Antoine Labarraque had trained as a pharmacist and was familiar with *eau de javel*, a solution of sodium hypochlorite first formulated by the French chemist Claude Berthollet in 1789. He knew that in addition to bleaching fabrics, the solution also imparted a fresh odor. Stimulated by the offer of a prize, Labarraque investigated using Javel in the processing of intestines and discovered that not only did it reduce the smell, the hypochlorite also facilitated the separation of the intestine's components. He won the prize and went on to study the use of various hypochlorite solutions to deodorize toilets, abattoirs, anatomical theaters, morgues, stables and hospitals.

Since disease at the time was thought to be spread by noxious fumes of bad air, or miasmas, Labarraque began to recommend that smelly wounds be rinsed with Javel and that the solution be used to clean places where people had been afflicted with cholera. He also suggested that doctors breathe Javel's vapors and wash their hands with a hypochlorite solution. This was long before John Snow in England connected cholera to drinking polluted water, and way before Ignaz Semmelweis made the link between a lack of hand washing and childbed fever. It wasn't until 1847 that Semmelweis theorized that doctors who attended to patients after coming straight from dissection rooms were causing disease by transferring smelly "cadaveric particles."

Semmelweis knew that Labarraque's solutions reduced the smell of decay and advocated their use for hand washing, finding hypochlorite worked better than soap. Since it was he who actually documented that childbed fever risk was reduced by hand washing, Semmelweis gets credit for the discovery, even though Labarraque had advocated hand washing with Javel some twenty years earlier. Of course, neither Labarraque nor Semmelweis knew about bacteria; the germ theory of disease would not be cemented by Louis Pasteur and Robert Koch for another couple of decades. And it wasn't until 1894 that German chemist Moritz Traube demonstrated that chlorine solutions killed bacteria, precipitating the first addition of chlorine to a municipal water system in Middelkerke, Belgium. Chlorination of drinking water turned into one of the greatest scientific advances in history. And to think that it all can be traced back to a need to deodorize "gut factories"!

Those gut factories also played a large part in the development of the impressive airships that launched the age of air travel in the first half of the twentieth century. These cigar-shaped masterpieces of engineering, named after their German pioneer Count Ferdinand von Zeppelin, featured a canvas-covered aluminum framework hull that housed up to twenty giant bags filled with hydrogen. Obviously these bags had to be extremely light, strong and leakproof. Goldbeater's skin fit the bill perfectly! One of the amazing properties of the skin is that wet separate sheets can be welded together by overlapping and gentle rubbing to form a seamless, leakproof joint. The zeppelins were gigantic, requiring incredible amounts of goldbeater's skin to hold them aloft. More than a million cows contributed their intestines to the most famous dirigible of them all, the *Hindenburg*.

Sixty-one people survived the horrific crash of this airship in 1937 as it was about to land in Lakehurst, New Jersey. Many undoubtedly had their wounds sutured with catgut, which has nothing to do with cats. The term likely originates from "kit," an old name for a fiddle, the strings of which, like suture thread, were made from animal gut, usually that of sheep. Catgut sutures, being composed of biological material, eventually decompose, eliminating the need for removal — a property discovered by Joseph Lister, who is mainly remembered for his introduction of phenol to reduce infections during surgery. He also pioneered the idea of washing catgut in a phenol solution after noting that raw catgut introduced infections into wounds.

Now, should you ever be asked to find a link between King Tut's golden sarcophagus, the *Hindenburg*, dissolving sutures and King Louis XVIII's body odor, you'll be prepared to give a gutsy answer.

THERE'S COPROLITES IN THEM THAR HILLS

You've undoubtedly heard of the California gold rush that started in 1848. But any mention of the "coprolite rush" in southern England that took place around the same time is likely to draw blank stares. Yet digging for coprolites in the fields of Bedfordshire and Cambridge probably had a greater impact on society than panning for gold in California.

So, what are coprolites? The term, from the Greek "*kopros*" for "dung," and "*lithos*," for "stone," was coined in 1829 by the Reverend William Buckland, first professor of geology and minerology at the University of Oxford. Buckland had identified some mysterious stones he discovered on a geological excursion as fossilized animal droppings. Before long it became clear that this was not an isolated find: the land was full of thick seams of the fossilized remains of animals that had succumbed to a major rise in sea level millennia earlier. Fossilization occurs when the empty spaces within a buried organism fill with water rich in minerals that then combine with the organism's natural chemicals to form a solid deposit. In the case of coprolites, that deposit is calcium phosphate. And that is one important chemical. Indeed, when it comes to feeding the world, it is essential!

Phosphorus is a vital element for all living organisms. It is a component of biomolecules such as DNA, the classic "blueprint" for life, as well as of ATP, the molecule that transports chemical energy in cells. Bones and tooth enamel are made mostly of hydroxyapatatite, which is a complex form of calcium phosphate. Where does all the phosphorus we need come from? Basically from plants that we either eat directly or eat indirectly through meat. And where do plants get their phosphorus? From the soil.

The connection between the composition of soil and a plant's nutrient needs was first formulated in a systematic fashion by the noted German chemist Justus von Liebig early in the nineteenth century. Liebig was concerned that farm production could not keep pace with the growth of the population unless crop yields could be increased. Being a master at chemical analysis, Liebig managed to determine the chemical composition of many plants and concluded that for growth they had to absorb a number of elements from the soil, with phosphorus being the crucial limiting factor. It didn't matter if there were enough of the other nutrients, an insufficient supply of phosphorus would stunt growth.

In his classic 1840 work entitled *Organic Chemistry and Its Applications to Agriculture and Physiology*, Liebig provided a formula for the combination of minerals he believed should be added to soil to replenish its nutrients, emphasizing the importance of "rock phosphate," a naturally occurring form of calcium phosphate. This was really the beginning of what might be called scientific farming, a concept that appealed to John Bennet Lawes, a Hertfordshire landowner who decided to test Liebig's theory. And who better to help with the experiment than Henry Gilbert, a chemist who had trained under Liebig in Germany?

Even before he heard of Liebig's work, Lawes was aware of the possibility of fortifying the soil with calcium phosphate. He knew that some time in the early 1800s, farmers had discovered that waste bone shavings dumped by Sheffield knife manufacturers made the soil more fertile. This had actually precipitated a frantic search for crushed bones, with battlefields being scoured and mummified cats from Egyptian pyramids and skeletons from Sicilian catacombs being imported into England. Lawes knew that bones were essentially a source of calcium phosphate, so this meshed with Liebig's theory of fertilization.

But when he tested "rock phosphate," he found it to be relatively insoluble, making the absorption of phosphorus by a plant's roots inefficient. He wondered whether the solubility of calcium phosphate could somehow be increased and eventually discovered that treating the bones with sulfuric acid did exactly that. "Super phosphate of lime," as the novel substance was christened, turned out to be amazing, making Lawes's turnips grow at an unprecedented rate! It was now time to capitalize on the discovery with the launch of the Lawes Chemical Manure Company.

Liebig did not take the challenge to his authority by a "mere farmer" well and proceeded to sue Lawes, claiming that he had actually come up with the idea of dissolving calcium phosphate in sulfuric acid first. Lawes won the suit, and with that William Buckland's discovery of coprolites took on new importance. Here was a ready, easily available source of calcium phosphate. By the 1860s, digging for coprolites became a huge industry in Bedfordshire and Cambridgeshire, particularly around the town of Shillington.

Harvesting coprolites was hard work, as was separating the fecal remains from clay in wash mills. And treating the residual powder with sulfuric acid was a nasty job that resulted in many accidents. No wonder that wages for coprolite workers were higher than for agricultural laborers! The men spent money freely, especially in the numerous pubs that opened. Coprolite miners turned out to be a huge boost to the local economy with blacksmiths, carpenters, boot makers and particularly brewers benefiting. But there were also reports of drunkenness, theft, rape and assault.

The boon and the coprolite industry itself skittered to a halt at the end of the century as Europe became flooded with cheap fertilizer made from phosphate deposits that had been discovered in America. Today, there is concern that these deposits will soon run out and that the world will face a fertilizer crisis. Since humans excrete a lot of phosphorus, there are major efforts being made to recover phosphorus compounds through sewage treatment. Under the right conditions, magnesium ammonium phosphate, or struvite, an effective fertilizer, can be extracted from waste water. Struvite may be this century's coprolite!

Today there is virtually no vestige of what at one time was a thriving industry in England, although there still is a Coprolite Street in Ipswich. Actually, one other curious reminder remains. Before 1881, the name of Shillington was actually "Shitlington," but the spelling was supposedly changed in order to prevent a shock to Queen Victoria's ears should her majesty show some interest in the coprolite industry.

SKELETONS IN THE CLOSET

Where would you go to see the largest illuminated advertising sign in the world? Times Square? Piccadilly Circus? No. You would have to hop over to Leverkusen, Germany, to be amazed by the 120-meter-high, 51-meter-diameter aspirin tablet atop the Bayer Company's headquarters. The 1,710 light bulbs that make up the famous Bayer cross on the giant tablet are meant to call attention to the shining history of the Bayer Company. But there is reason to take a dim view of some aspects of that history.

There's no question that Bayer has a very colorful past, beginning appropriately enough with synthetic dyes, its first commercial product. Friedrich Bayer was a paint and dye salesman before partnering with master dyer Johann Friedrich Weskott to establish a factory in 1863, hoping to get in on the mushrooming synthetic-dye business. Historically, dyes were isolated from plant sources such as indigo and madder, from sea snails or from insects like kermes or cochineal. All that changed with William Henry Perkin's chance discovery of synthetic dyes in 1856. His accidental synthesis of mauve while attempting to make quinine from coal tar launched the synthetic dye industry and precipitated widespread research into coal tar dyes by an array of competitors.

A rainbow of synthetic colors soon appeared, with German companies such as Bayer taking the lead. Competition was fierce, forcing the companies to diversify. If dyes could be produced synthetically in the laboratory, why not pharmaceutical products? Why not indeed. By 1883 Hoechst had produced antipyrine, the first fever- and pain-reducing drug, and Bayer waded into the waters with phenacetin in 1888. Then in 1898 came Bayer's big breakthrough with aspirin, marking the beginning of a long string of spectacular achievements as well as disturbing controversies.

First, a bit of history. Contrary to many popular accounts, aspirin is not produced from the bark of the willow tree. The starting material for the chemical synthesis of aspirin is benzene, derived from petroleum. This is then converted to phenol, which in turn is converted to salicylic acid, which is then converted to acetylsalicylic acid, or ASA, which we commonly know as aspirin. While aspirin is not made from willow bark, there is a connection. The bark of the white willow, as well as several other plants such as myrtle, contain various compounds that have a chemical similarity to aspirin and which, like aspirin, have an effect on pain, fever, inflammation and blood clotting. The Ebers Papyrus, an Egyptian medical text dating back to the sixteenth century BC, describes the use of the bark of the willow tree as well as myrtle to treat pain, fever and conditions that today we would describe as inflammatory. In the fifth century BC, Hippocrates recommended willow bark preparations to reduce fever and ease the pain of childbirth.

Recipes for various willow extracts made their way into pharmacopoeias around the world, but it wasn't until the eighteenth century that the first scientific study of willow bark was carried out. Credit for this study, which certainly did not have the rigor we expect from studies today, goes to the English clergyman Edward Stone. A common belief at the time was that cures for diseases were to be found near the cause of the disease. This was known as the doctrine of signatures and had a quasi-religious bent. The notion was that God had given humans clues about finding remedies for diseases. Since fevers were more common around brackish waters, that's where Stone searched for a clue.

Of course the reason that fevers were to be found in such areas was because such waters served as a breeding ground for mosquitoes, which transmitted malaria — a connection not known at the time. Stone tasted the bark of the willow tree and found it bitter, reminiscent of the taste of cinchona bark, the only known effective treatment for malaria. Cinchona was later found to contain quinine, and it was the bitter taste of this compound that suggested to Stone that he may have discovered another malaria cure. He collected and powdered some willow bark, which he then went on to systematically test on people who complained of fever and aches. While it did relieve symptoms, it did not, like quinine, cure the disease. Still, willow bark gained popularity as an inexpensive substitute for cinchona bark.

The nineteenth century ushered in the modern era of chemistry, with its powerful tools of isolating, identifying and synthesizing compounds. One of the main areas of interest was the isolation of physiologically active substances from natural materials. By the middle of the century salicin and salicylic acid had been isolated from willow bark and were widely marketed for the reduction of pain, fever and inflammation. Attempts were also made to produce synthetic derivatives of the compounds found in willow bark and one of these was acetylsalicylic acid, synthesized by Charles Frédéric Gerhardt in 1853. Gerhardt did not investigate the medical properties of his novel compound. It remained for Felix Hoffmann, a chemist working for the Bayer company, to put the final punctuation mark on the drug that would become known as aspirin.

One of the problems with natural salicylates was that they commonly cause gastric distress. Hoffmann's father apparently suffered from arthritis and got relief from natural salicylates, but at the expense of gastric pain. Hoffmann thought that perhaps a synthetic derivative of these salicylates might retain the therapeutic properties and eliminate the gastric problems. He searched the literature and came upon Gerhardt's synthesis of acetylsalicylic acid, which he used to produce enough material for testing. The compound relieved fever, pain and inflammation, and while it did not eliminate gastric problems, it did reduce them relative to salicin and salicylic acid.

Bayer coined the term aspirin from the Latin "*a*" for "from" and "*spirsaure*," the German for "salicylic acid." That term derived from *Spiraea ulmaria*, the botanical name for meadowsweet, which is rich in salicylic acid. This account of the discovery of aspirin, repeated in numerous articles and texts, was first publicized by Bayer in 1934, a year after the Nazis came to power. Arthur Eichengrün, a German-Jewish chemist who was Hoffmann's supervisor at the time of the discovery, later claimed in 1949 that the research was actually his idea and that Hoffmann just followed directions and didn't even know why he was asked to synthesize the compound. Eichengrün alleged that the Hoffmann story was released to ensure that a Jewish scientist would not be credited with the discovery. Bayer has denied this claim, but medicinal chemist Dr. Walter Sneader, after researching the issue extensively, has concluded that Eichengrün's account was credible and that the Nazi propaganda machine was instrumental in the ethnic cleansing of aspirin's history.

Why did Eichengrün wait until 1949 to publish his account? Certainly, Nazi Germany was no place to make claims about Jewish discoveries. Although Eichengrün's story only received attention after the war, he had originally described the events in a letter to Bayer he somehow managed to write from a concentration camp where he had been interned by the Nazis at the age of seventy-six. In the letter, now in the company's archives, he asked for Bayer's help. The letter was ignored, as is the whole controversy in the "Bayer History Fact Sheet" on the company's website.

During the World War II, Bayer merged with other chemical companies to form IG Farbenindustrie AG. Not only did this company produce the notorious Zyklon B used in the gas chambers as well as the nerve gas sarin, it used slave labor in its factories and played a large role in the horrific drug experiments on human subjects in Auschwitz. The history fact sheet describes these events thus: "As Germany's most important chemical company, IG Farbenindustrie became involved in events during the Third Reich." Yeah, "involved."

Curiously, Bayer's marketing of heroin from 1898 to 1913 also fails to appear in its official history. This compound was also synthesized by Hoffmann at virtually the same time as aspirin. He was trying to make codeine from morphine when he hit upon heroin, which had actually been first synthesized back in 1874 by British chemist Charles Romley Alder Wright. Bayer's tests showed that subjects taking the drug during testing felt heroic and brave, hence the name. Heroin turned out to be an effective cough suppressant, in great demand at a time when tuberculosis was running rampant. It was widely marketed by Bayer as "non-addictive," although that turned out not to be the case. No great surprise here, since in the body heroin is quickly converted to morphine. Whether the company actually knew this while it was promoting heroin as non-addictive is still a matter of debate. Also missing from the fact sheet is any mention of Baycol, the anticholesterol drug that was withdrawn in 2001 because it allegedly caused kidney failure and death in fifty-two patients.

Of course there have also been discoveries that have significantly enhanced our lives. The first antimicrobial drug, sulfonamide, was discovered by Bayer's Gerhard Domagk in 1932. Otto Bayer invented polyurethane in 1937, Hermann Schnell introduced polycarbonate plastics in 1953 and then of course, there's aspirin. Today, Bayer researchers are working on new pharmaceuticals, novel methods of crop protection and ways to use carbon dioxide in plastic production. Indeed they are justifying the company's current slogan, "Science for a Better Life." And that goes for birds too. For a period in the spring and fall each year, the lights on the giant Bayer cross sign are switched off at night to ensure that migrating birds can reach their breeding grounds. And in a move toward "greenness," the 1,710 conventional 40-watt light bulbs have been replaced by innovative light-emitting diodes (LEDs), resulting in an 80 percent reduction in energy use.

Yes, there are skeletons in Bayer's closet. As in those of most big corporations. (And most people's.) Shedding light into the dark nooks is important, since where we go in the future depends on what we have learned from the past. But we also have to ensure that as we illuminate the past we do not cast an unfair shadow on the present.

BLANCMANGE!

I recently learned a new word: "Blancmange." Turns out, it's a sweet dessert made with cream, sugar and almonds, thickened with cornstarch or gelatin. Strawberry-flavored gelatin gives it a pink tinge, which is how I came to hear of this treat. "It's a blancmange," outraged citizens sniggered when they got their first glimpse of the freshly painted Angel Hotel in Lavenham, Suffolk. Celebrity chef Marco Pierre White had purchased the hotel and repainted it pink, thinking it would fit in nicely with the rosy palette of colors that characterize Suffolk towns. But it turned out to be the wrong pink. It was not the famous "Suffolk Pink"!

"Suffolk Pink" takes us back to medieval times when Suffolk was the center of the textile industry in England, which meant that dyers also had a flourishing practice in the area. Plants in the countryside provided an array of pigments, including yellows from cow parsley, reds from the roots of lady's bedstraw, purples from the skin of damson plums and pinks from elderberries. Most houses at the time were whitewashed, so it comes as no surprise that some folks got the idea of jazzing things up a bit by adding a little color. But this required a lot of plant extract, so the search was on for something that was more practical. As it turned out, the blood of pigs or oxen fit the bill! And so the Suffolk Pink tradition featuring "bloody whitewash" was born.

Whitewash itself presents some interesting chemistry. Remember when Aunt Polly asked Tom Sawyer to whitewash a fence, a task he was not particularly keen to perform? Some quick thinking allowed Tom to sit in the shade while others did the work for him. Cleverly, he let on that this difficult task had been entrusted to him, and him alone, because others could not do the job properly. With that gauntlet thrown down, he suckered passers-by into begging him for a shot at whitewashing.

Tom held out until they could come up with adequate payment, eventually netting him an apple, a bunch of marbles and even a dead rat that could be twirled around the head with a string. Neither Tom nor his friends realized that they were actually engaging in a fascinating chemistry experiment.

The white layer that deposits on whitewashed surfaces is calcium carbonate. But that is not the chemical that is applied. Whitewash is actually a solution of calcium hydroxide, which after application reacts with carbon dioxide from the air to form calcium carbonate. Interestingly, this is a "dust to dust" sort of phenomenon. Calcium hydroxide is made by combining calcium oxide (lime) with water, and the calcium oxide in turn is made by heating limestone made of calcium carbonate to drive off carbon dioxide. This whitewash reaction is often performed in high-school chemistry labs to demonstrate the presence of carbon dioxide in exhaled breath. Blowing into a test tube filled with limewater, which is just calcium hydroxide dissolved in water, causes a precipitate of calcium carbonate to form.

Blood and plant pigments are not the only coloring agents that can be added to whitewash. Blue vitriol, or copper sulfate pentahydrate, produces a striking blue. And that now takes us from Suffolk in England to Jodhpur, the "Blue City" of India, where in one area of the city virtually all the houses are painted a light sky blue. Why this is so has been a matter of some mystery. The most widely accepted theory seems to be the association in India of blue with royalty and power. The highest caste of Hindus, the Brahmins, have traditionally lived in the blue district of Jodhpur and supposedly wanted to distinguish their homes from those of the lower-class Hindus. Brahmins, being more affluent, were able to afford the more expensive copper sulfate–based whitewash.

There may, however, be a more practical explanation for the

blue houses. Copper sulfate protects walls from termites, which plague the area. It also has mold-inhibiting properties, a decided plus for painted walls. To this day, copper sulfate is used as a fungicide, especially on grapes, in a product called a Bordeaux mixture. Since it occurs in nature, crystallized on walls of caves and mines, it is classified as an "organic" fungicide although it is actually produced industrially by reacting copper oxide, which is mined, with sulfuric acid. Of course, copper sulfate is copper sulfate no matter how it is produced. It is promoted by the organic movement as "natural" and therefore safe, despite the fact that copper sulfate has significant toxicity, particularly for fish.

For me, copper sulfate brings back some memories. I remember opening my first Gilbert chemistry set and being struck by a little bottle with the stunning blue crystals. That was back in the days when chemistry sets actually contained chemicals. Not like some of the absurd kits available today that advertise themselves as containing "everyday products without chemicals." (I wonder what everyday product they have in mind that contains no chemicals.) In any case, my chemistry set did contain chemicals, as does everything in the world, and I remember the excitement of dipping an iron nail into a solution of copper sulfate and watching it slowly being plated with a layer of shiny copper.

Now back to Lavenham. Like other Suffolk towns, it has a "village plan," advising homeowners of the colors that are in keeping with protecting the town's medieval heritage. It is not expected that people will use pig's blood, but commercial pink paints are to have the right hue. This is where Marco ran afoul of heritage sensitivities and raised the ire of the citizens. Truth be told, he was already in some hot water because of his decision not to serve lager or Guinness in the Angel's pub, commenting that this would attract the "wrong sort of people" to the establishment. That did not go down well, given that the Angel was originally built as a tavern in 1420. The uproar was great enough to have Marco agree to repaint the hotel, putting him back in the pink with the citizens of Lavenham. I don't know if the pub serves blancmange, but a pink version would be a clever marketing touch.

TOILET SCIENCE

International Toilet Day is celebrated every year on November 19. That may sound funny, but it is no joke. It is a time to contemplate what we have and others don't. As we sit in privacy on our comfortable flush toilets today, it is hard to imagine that a scant 200 years ago sewage disposal meant emptying chamber pots into the nearest convenient place, which was often the street. If you were out for a walk in Britain in the eighteenth century and heard the cry "gardy-loo," you had better scamper across the street because the contents of a chamber pot were set to be hurled your way from a window. The expression derives from the French *regardez l'eau* and was commonly heard as chambermaids carried out their duties. Some even suggest that the custom of a gentleman walking on the outside when accompanying a lady can be traced to the desire to protect the fair sex from the trajectory of the chamber pot's contents.

What may be even harder to imagine than the sidestepping of flying fecal matter is that roughly a third of the world's population today cannot easily sidestep the problems associated with exposure to untreated sewage because of lack of access to a toilet. As a consequence, diarrheal disease is rampant, killing more children than AIDS, malaria and measles combined. In developing countries, a child dies every twenty seconds as a result of poor hygiene. Mahatma Gandhi recognized the problem when he proclaimed in 1925 that "sanitation is more important than independence."

The invention of the flush toilet and the introduction of plumbing for sewage disposal mark two of the most significant advances in history. Let's get one of the toilet myths out of the way right away. Contrary to numerous popular accounts, Thomas Crapper did not invent the flush toilet! It is easy to see how connecting his name with the invention would make for a compelling tale, but what we actually have here is a prime example of the classic journalistic foible, "a story that is too good to check."

Almost all accounts of the Crapper saga claim that a 1969 book by Wallace Reyburn, cleverly titled *Flushed with Pride: The Story of Thomas Crapper*, establishes Crapper as the inventor of the flush toilet. Reyburn actually says no such thing. The book is an entertaining celebration of the life and times of Crapper, the man who "revolutionized the nations' water closets." Indeed, that he did do. But flush toilets were around long before Thomas Crapper ever got into the game in the nineteenth century.

The first flush toilet appeared as early as 1700 BC. The Palace of Knossos on the island of Crete, built around that time, featured a toilet with an overhanging cistern that dispensed water when a plug was removed. Curiously it would take another 3,000 years until the next step in flushing technology was taken by Sir John Harrington, godson of Queen Elizabeth I. In 1596, Harrington installed a "water closet" in the Royal Palace that featured a pipe fitted with a valve connected to a raised water tank. Opening the valve released the water that would carry waste into a cesspool. Apparently the Queen was not overly pleased with the invention because odors from the cesspool wafted up into the Royal powder room. It would take another couple of centuries before this problem was addressed.

The first patent for a flushing toilet designed to keep sewer gases from seeping back was issued to Alexander Cummings in 1775. Cummings designed a system that allowed some water to remain in the bowl after each flush, preventing the backflow of odors. Joseph Bramah attempted to improve upon this system with a sophisticated valve that was supposed to seal the waste pipe after each flush. While it didn't work perfectly, Bramah's toilet was introduced at just the right time because London was beginning to install sewage systems. Some 6,000 Bramah toilets soon dotted the city's landscape. And then about a hundred years later, along came Thomas Crapper.

In 1861, Crapper's plumbing company opened for business in London. The time was ripe for the sale of plumbing supplies because the need for proper sanitation was being firmly established. A public report issued in the city of Leeds claimed a significantly higher death rate among children who lived in "dirty" streets where sewage flowed openly. And in 1854, physician John Snow had pinpointed the homes in London where someone had contracted cholera during an epidemic and traced the problem to water contaminated with sewage being dispensed from a pump in Broad Street. The need to flush away problems associated with sewage was becoming clear.

There is no question that Crapper made significant improvements in toilet technology. He invented a pull-chain system for flushing and an air-tight seal between the toilet and the floor. Crapper was also responsible for installing plumbing at Westminster Abbey, where to this day visitors can view the manhole covers clearly displaying the name "T. Crapper & Co." What he was not responsible for was the introduction of the word "crap" into our vocabulary. The use of the term to mean "refuse" predates Crapper by several centuries.

It is virtually impossible to attribute the numerous improvements in toilet technology since Crapper's time to individuals. There are patents galore for eliminating overflow, reducing water usage, curbing noise, improving waste removal from the side of the bowl, devices to alert night time users if the seat is up and gimmicks to encourage men to aim properly. And the future may belong to toilets equipped with biosensors that automatically monitor urine and feces for health indicators such as sugar and blood. But for now, just think of the amazing technology that allows for the removal of the roughly 200 grams of poo we deposit per person per day. That's a stunning 600,000 kilos in a city of three million!

So next time November 19 rolls around, as you get comfy on your high-tech toilet, ready to flush away the remnants of a scrumptious meal, a roll of soft toilet paper and fragrant soap by your side, give a thought to how we can help those unlucky enough to have been born in a place where "gardy-loo" still rings true.

MUSIC AND CHEMISTRY, LIVING IN PERFECT HARMONY

Quick now, name some famous scientists. Einstein surely comes to mind, then maybe Newton, Marie Curie, Galileo, Stephen Hawking. Not much of a problem. Next, name some famous composers. I suspect you'll quickly reel off the names of Mozart, Beethoven, Brahms, Tchaikovsky and Liszt, along with a host of others. No problem at all. But now name a scientist who also made major contributions to the world of music! Most people will draw a blank, but a few will come up with Alexander Borodin. Who, you ask? Alexander Borodin, chemist and Tony Award winner, that's who!

Unfortunately Borodin, the composer of much of the music for *Kismet*, the winner of the 1954 Tony for Best Musical on Broadway, was not around to receive the award. That's because he had been dead for sixty-seven years! The Russian composer/scientist had a fascinating dual career, characterized by dashes between the laboratory and the piano, with success springing from both. Music aficionados know him as a celebrated composer, with nary an idea that he was a highly regarded professor of chemistry. Chemists know of his exploits in the lab, but are unlikely to be familiar with his symphonies.

Alexander Borodin should not have been a Borodin. He should have been a Gedeoneshvilli, the name of his father, a Russian prince. The problem was that Alexander's mother was the prince's maid and marriage was out of the question. As was common practice in those days for children born out of wedlock, a serf was appointed as the "father" and his name bestowed on the newborn. Porfiry Borodin, the prince's valet, was the chosen one, with his wife being listed as the boy's mother. Later, Alex's real mom, who would always be referred to as "Aunt

Mimi," would marry a retired army doctor, allowing the boy to grow up in a privileged household. He learned French, German, Italian and English and mastered the cello, flute and piano. At the age of nine, Alex promptly fell in love with a young girl named Elena, for whom he composed his first musical work.

But science also captured the boy's imagination, particularly the chemistry of fireworks. Before long, "Auntie's" apartment was filled with flasks, beakers, jars of crystals and an assortment of chemical smells. Borodin loved chemistry and wanted to study it formally, but at the time in Russia there were no degrees being granted in the subject. It was, however, part of the medical curriculum, so if you wanted to study chemistry, you had to enroll in medical school, which Borodin did. He graduated in 1858 with his thesis, a requirement for a degree, "On the Analogy of Arsenic Acid with Phosphoric Acid in Chemical and Toxicological Behaviour." With the defense of that thesis, Borodin was granted the degree of medical doctor, which was somewhat of a paradox, seeing that he was prone to fainting at the sight of blood! Never would he practice medicine, which as it turned out, would prove to be a boon both for chemistry and music.

Apparently the new graduate impressed the Russian government enough to send him to Europe for four years to absorb the latest developments in science. In Heidelberg he encountered countryman Dmitri Mendeleev, the "Father of the Periodic Table," who provided further chemical inspiration. So did Emil Erlenmeyer, the German chemist probably best known for the conical flask he designed, a standard piece of laboratory equipment. From Erlenmeyer, Borodin learned about the chemistry of a family of compounds known as the aldehydes, which put him on a path leading to the discovery of a classic reaction known as the aldol condensation, often incorrectly attributed to the French chemist Charles-Adolphe Wurtz.

The aldol condensation has numerous applications, including in the synthesis of Lipitor, the famous cholesterol lowering drug, as well as in the production of cinnamaldehyde, the compound that lends cinnamon flavor to buns and apple pie. As famed chemistry historian George Kauffman explains, Borodin also developed methods for adding fluorine or bromine to organic compounds, for determining the presence of urea and for analyzing the contents of tea and mineral waters.

After his sojourn through Europe, Borodin returned to Russia to take up a professorship at his alma mater, the Academy of Medicine. By all accounts he was a great teacher, dedicated to his students, but music was always on his mind. Between lectures he was known to jot down a catchy tune! Eventually Borodin would compose symphonies, chamber music, songs, piano works and an unfinished opera, *Prince Igor*. His highly original melodies often had an Islamic flavor, as is evident in *Kismet*.

Borodin deserves yet another laurel. One day he heard a young Russian pianist play at a concert and realized she had absolute pitch. Instant love! Ekaterina Protopopova and Borodin were soon married, and Borodin became a passionate supporter of his wife's crusade for women's rights, eventually becoming one of the founders of the School of Medicine for Women in St. Petersburg. To this day, women make up a large proportion of doctors in Russia.

Repeatedly throughout his career, Borodin expressed his dismay at having a lack of time. When he was in the lab, he yearned to compose music. When he sat at his piano, his thoughts drifted to his chemical experiments. His musician colleagues urged him to give up science, while his chemical mentor, Professor Nikolai Zinin, admonished him for wasting too much time thinking about music, telling him, "You cannot hunt two hares at the same time."

Sadly, at the age of fifty-three, Alexander Borodin ran out of time altogether, departing this world, probably as he would have liked, with musical accompaniment, surrounded by his scientific colleagues. He had organized a fancy dress ball at his home to celebrate the achievements of the professors of the Academy of Medicine, when in the midst of an enthusiastic dance he collapsed and died. Who knows what else Alexander Borodin would have accomplished had he lived to a ripe old age? Perhaps there would have been another award alongside the Tony. Maybe a Nobel?

CHLOROFORM!

The experiment, James Simpson decided, would go ahead even though the rabbits died. And that decision was destined to have such a huge impact on medicine that when Simpson passed away in 1870 more than 100,000 grateful Scots lined the streets of Edinburgh to pay their respects as his funeral cortege passed by. Many had undergone surgery or had given birth to children painlessly thanks to Simpson's great discovery: chloroform!

Not many students graduate with a medical degree at age twenty, but that is just what James Simpson did. In the early years of the nineteenth century, doctors didn't have many tools at their disposal, and young Simpson was particularly disturbed by the suffering he witnessed in the surgical theater. He once remarked, "The man laid on an operating table in one of our surgical hospitals is exposed to more chances of death than the English soldier on the field of Waterloo." No surprise then that when word came from America of the discovery of ether as an anesthetic, Simpson was quick to jump on the bandwagon. But there were problems with ether. It was flammable, difficult to administer and often made patients sick. However, if ether could put people to sleep, Simpson reasoned, there had to be other chemicals that could do the job as well. And who better to ask for suggestions than Scottish chemist Lyon Playfair who had trained under famed German professor Justus von Liebig.

Playfair was no expert in putting people to sleep, but told Simpson about a sweet-smelling volatile liquid he had seen Liebig prepare back in 1832. It just so happened, Playfair went on, that one of his assistants had recently made some, and Simpson was welcome to give it a try with one proviso. The experiment would first have to be tried on two rabbits. Playfair, it seems, didn't want to be responsible for any harm that could befall the eager Simpson, who was already establishing a name as a caring, skilled physician.

Without further ado, the rabbits were exposed to the vapors of chloroform and promptly fell asleep. When they awakened with no apparent side effects, an exuberant Simpson grabbed a bottle of chloroform and made plans with friends Drs. George Keith and Matthew Duncan for a little chloroform party the next day. The three had already been into sniffing a variety of chemicals with hopes of finding an improved anesthetic, but Keith and Duncan suggested it would be a good idea to check on the rabbits in the morning for any lingering effects. Well, there were lingering effects, all right. The rabbits were dead!

Why this did not deter the trio from experimenting with chloroform isn't clear. Perhaps it was their drive to become medical pioneers, or maybe they had enjoyed the effects produced by some of the chemicals they had previously inhaled. In any case, on the evening of November 4, 1847, Keith became the first guinea pig, inhaling a good dose of chloroform. Within a couple of minutes, he was under the table. Without waiting to see their colleague's fate, Duncan and Simpson followed suit. After some initial hilarity, they also passed out. On awakening, Simpson declared, "This is far stronger and better than ether," and predicted chloroform would "turn the world upside down."

The young doctor was so impressed that he immediately hired a chemist to prepare a fresh supply of chloroform, which in those days was made by reacting acetone with chlorine. Four days after his sniffing binge, Simpson chloroformed a woman who twenty-five minutes later gave birth. Within a month, Simpson had used chloroform successfully on more than fifty patients! Still, as with any drug, there were risks, including death. In 1848, young Hannah Green died, probably due to improper administration of the anesthetic for the removal of an infected toenail. Soon after this tragedy, Dr. John Snow developed an inhaler that regulated the dosage of chloroform and reduced the risk of such deaths. Simpson was then able to quiet those who still deemed chloroform to be too dangerous by keeping careful statistics of successes and side effects. When detractors claimed that chloroform was unnatural, Simpson replied, "So are railway trains, carriages and steamboats."

Religious opposition, however, was harder to overcome. The faithful argued that pain relief during childbirth was unholy because according to the scriptures women were destined to be punished for Eve's original sin. Tempting Adam with the fruit of the tree of knowledge, it seems, was not a good idea. Critics pointed to a passage in Genesis where God tells Eve, "I will increase your pains in childbearing; with pain you will give birth to children." Simpson cleverly retorted that anesthesia was actually inspired by God, who can be regarded as the world's first anesthetist. He had his own biblical reference: "So the Lord God caused the man to fall into a deep sleep; and while he was sleeping, he took one of man's ribs and closed up the place with flesh."

Still the battle between the pro- and anti-chloroform forces raged for several years, finally abating in 1853 when Simpson recommended chloroform to his most famous patient, Queen Victoria. After giving birth to Prince Leopold, the Queen declared she was "much gratified with the effect of the chloroform." If chloroform was good enough for her Majesty, it was good enough for her subjects!

As the twentieth century rolled in, chloroform began to show a notorious side as well. Criminals had taken to incapacitating victims by clamping a rag soaked with the liquid over their mouth. Then in a highly publicized case in 1901, American businessman William Marsh Rice, whose fortune founded Rice

University, was killed by his valet who aimed to get his hands on Rice's assets. The murder weapon was chloroform!

More problems cropped up by the 1930s, with chloroform being linked with liver problems and irregular heartbeats that sometimes proved fatal. Finally, as new sleep-inducing agents such as isoflurane and desflurane were introduced, the use of chloroform as an anesthetic was relegated to the junk heap of history. But in its time, that "junk" was responsible for alleviating a great deal of suffering. In commemoration, an impressive bronze sculpture of Simpson now dominates Princes Street Gardens in Edinburgh, with the inscription, "Pioneer of Anaesthesia."

BREAKING THE CODE

Winston Churchill said Alan Turing's breaking of German codes for secret messages shortened the war by two years. Computer scientists say he formulated the philosophical principles that power every computer and smartphone today. Researchers in artificial intelligence label him as the founder of their field. Chemists say he predicted the existence of reactions that change color in a periodic fashion before these had ever been observed. Biologists say his theories of morphogenesis, the biological process that causes an organism to develop its shape, explain phenomena such as the stripes of a zebra by predicting the effects of the diffusion of two different chemical signals, one activating and one deactivating growth. The British government said he was a homosexual who needed to be chemically castrated. The police said his death was due to suicide, probably by consuming a cyanide-laced apple. Steve Jobs said he wished the Apple logo had been devised in his honor. And everyone says that Alan Turing was a computational genius. Few, however, know that his first scientific infatuation was with chemistry.

In 1924, at the age of twelve, young Turing got his hands on *Natural Wonders Every Child Should Know* by Edwin Brewster. He was totally taken by the book's discussion of plants that contained chemicals like strychnine and atropine, capable of killing or curing. He was also intrigued by how "carbon dioxide becomes in the blood ordinary cooking soda" and how "the blood carries the soda to the lungs, where it changes to carbon dioxide again, exactly as it does when, as cooking soda you add it to flour and use it to raise cake." He was overjoyed when for Christmas he was given a set of chemicals, crucibles and test tubes that he immediately put to use, extracting minute quantities of iodine from seaweed.

Knowing of his interest in iodine, Christopher Marcom, an older student with whom Alan had developed a friendship that bordered on infatuation, told him of an experiment involving solutions of iodates and sulfites that when mixed resulted in the formation of iodine. When starch was present in the solution, the formation of iodine was signaled by the sudden appearance of a deep blue color characteristic of a starch-iodine complex. Turing was intrigued by the concentration-dependent time delay in the appearance of the blue color and was convinced that the proper use of mathematics could predict the results. Indeed this little problem may have launched him into an area where his genius would be prominently displayed, namely mathematics and computer logic.

After receiving an undergraduate degree in mathematics from Cambridge, Turing obtained a PhD from Princeton, where he developed a strong interest in cryptography, the science of writing messages in secret code. When Britain declared war on Germany in 1939, Turing was immediately offered a position as a code breaker. The Germans had developed the Enigma, a machine that used a combination of mechanical and electrical effects to transform a message typed on a keyboard into what seemed like an undecipherable jumble of letters until the code was run through another Enigma unit. With colleague Gordon Welchman, Turing developed early computers, called "bombes," that were capable of breaking the code. The Allies now had the ability to decode messages sent to German U-boats about planned attacks on supply ships that carried supplies vital for the war effort.

After the war, Turing worked at the National Physical Laboratory designing computers and then in 1948 joined the Mathematics Department at the University of Manchester, where in 1950 he launched the concept of artificial intelligence with his paper that asked the question "Can Machines Think?" In it he also introduced the Turing Test, aimed at determining whether a human can detect if he is communicating via a keyboard with another human or with a machine. The emphasis is on how closely the answers to questions posed resemble typical human answers, not on whether the answers are correct. If at least 30 percent of judges believe that they are talking to a human when they are actually speaking with a computer, the computer is said to have passed the Turing test.

Now, for the first time ever, Eugene Goostman, a computer program created by Russian Vladimir Veselov and his team, has managed to fool the required number of judges into believing that they were actually speaking with a 13-year-old Ukrainian boy. The historic event took place at the renowned Royal Society in London, poignantly on the sixtieth anniversary of Turing's death. A death that is still mired in mystery.

In 1952, Turing's house was burgled and suspicion fell on a young male guest who had been staying with him. During the investigation, Turing's homosexuality was revealed and since this was a criminal offense at the time, he was charged with "gross indecency." The option, the judge explained, was either jail time or "chemical castration."

Back in the 1940s there had been speculation that homosexuality was caused by low levels of testosterone, the male sex hormone. But if anything, treatment with testosterone increased the sex drive. Researchers then turned to female hormones, hoping to counter the effects of testosterone, and soon discovered that these did indeed reduce libido. To avoid prison, Turing agreed to undergo a year-long treatment with the synthetic estrogen diethylstilbestrol (DES), a regimen that he apparently managed with no significant problems, carrying on as usual with his work.

We can only guess at what this brilliant mind could have accomplished had it not been put out of commission by cyanide in 1954. The official verdict was suicide, with speculation that a half-eaten apple found at the bedside had been used to deliver the poison. The apple was never tested, and Turing was in fact known to enjoy an apple at bedtime. Given that he had shown no signs of depression and had planned meetings for the next week, some contend that Turing accidentally poisoned himself with cyanide he was using in gold-plating experiments in his small apartment. And then there are the conspiracy theorists who maintain he was assassinated because authorities were worried that homosexuals who possessed sensitive information were a security threat to Britain.

In 2013, the Queen signed an official pardon for Turing's conviction for gross indecency. Indeed, the only indecency had been the appalling way in which one of the greatest minds of the twentieth century was treated.

SCREEN HOUDINI TIED UP WITH TWISTED FACTS

I'm an unabashed Houdini buff. While the fame of one of the greatest entertainers in history rests mostly on his escapes and magic, I've long been captivated by his battles with charlatans who claimed to be able to forge contact with the great beyond. Houdini insisted that they provide proper evidence, which was not forthcoming. Science is of course based on evidence, and for me Houdini was a great model of the proper pursuit of science. So is Sir Arthur Conan Doyle's Sherlock Holmes, another of my favorites. His classic line "it is a capital mistake to theorize before one has data because one insensibly begins to twist facts to suit theories, instead of theories to suit facts" has become my credo. Both Houdini and Holmes were adamant about relying on facts and not stretching the truth.

Holmes, I think, would have had a few bones to pick with the History Channel's much-ballyhooed mini-series about Houdini. After all, as Dr. Watson tells us in "A Study in Scarlet," Holmes had "a passion for definite and exact knowledge." And screenwriter Nicholas Meyer is certainly familiar with the detective's reliance on facts, having himself written *The Seven-Per-Cent Solution*, a highly acclaimed Holmes adventure. But in *Houdini*, Meyer exercised a fair bit of poetic license, which is curious because Houdini's real life is fascinating enough — it requires no alteration. The magician himself would likely have raised some objections, including one about the portrayal of his start in magic.

Young Ehrich Weiss, who would later pilfer the name of French magician Robert Houdin to become "Houdini," was totally taken by "Palingenesia," an illusion performed by traveling magician "Dr. Lynn." This is historically correct.

The curious name was fitting for an illusion that featured the dismembering of a "volunteer" and restoring him to perfect health, since "*palingenesis*" means "rebirth" or "recreation." "Palengenesia" was based on the strategic use of black fabric against a black background, as can be seen in an imaginative recreation by Neil Patrick Harris that is available on You Tube. Although some of the elements of Palengenesia were included in the television production, the main effect that so amazed Ehrich in the film was the classic sawing of a person in half. He could not possibly have seen this performed in 1884, since it was only in 1921 that P. T. Selbit in London and Horace Goldin in the U.S. first presented slightly different versions of the trick.

There are other issues as well. When Houdini meets Bess, whom he would marry after a whirlwind courtship, he impresses her by levitating a little paper bird in her hand. This fine trick is done with an "invisible thread," but the thin nylon strands only became available in the 1940s. As correctly presented, Houdini did vanish an elephant at New York's Hippodrome, but it was not done by draping a large cloth over the animal. Indeed, the film version would be impossible. The elephant actually disappeared in a large cabinet that was wheeled onto the stage. It took considerably more men to push the contraption off the stage after the elephant had "vanished." Why not depict this properly, as they did with Houdini's classic "walking through the wall" illusion? In this case, the producers even chose to reveal the actual method used: a trap door that allowed the magician to crawl under the wall.

Houdini never had a misadventure with the "Chinese water torture cell," forcing his assistant to use a sledgehammer to save his life. He never performed the "bullet catch" in front of the German Kaiser, and certainly was never punched in the stomach by Rasputin — whom he never even met. Margery, the medium whose tricks Houdini exposed, never attempted to seduce him in an attempt to safeguard her secrets. And the story about Houdini being a CIA agent is pure fantasy. But the most bizarre straying from facts involves the incident that may have cost the magician his life.

In 1926, Houdini was visited in his dressing room at the Palace Theatre in Montreal by one Gordon Whitehead, who asked about the magician's claim to be able to withstand any blow to the stomach. Houdini, who wasn't paying close attention, grunted a yes, but the blows came before he had a chance to prepare himself. Less than two weeks later, the most famous entertainer of the times would be dead from a ruptured appendix. In the film, the dressing room scene takes place in Detroit, not Montreal. It is true that after Montreal the troupe did travel to Detroit, where Houdini was to perform, but why not present the story as it really happened?

Nicholas Meyer did much better with the portrayal of the strange friendship between Sir Arthur Conan Doyle, the creator of Sherlock Holmes, and Harry Houdini. Conan Doyle, in a bizarre contrast to his scientifically minded detective, was a staunch believer in communicating with the spirit world. Houdini was highly critical of spiritualism and was the scourge of fake mediums who used magic tricks to convince the gullible that they had made contact with the dearly departed. Conan Doyle first met Houdini in 1920, and two years later when the writer came to America to lecture on spiritualism, Houdini invited his friend to the annual meeting of the Society of American Magicians. Conan Doyle was hesitant about accepting, fearing that the magicians would ridicule his spiritualist beliefs. But as it turned out, it was Sir Arthur who had something up his sleeve!

After the magicians had gotten through upstaging each other

with their newest tricks, it was the Scot's turn. He asked that the lights be turned down and a movie projector be brought in. The magicians' jaws dropped when they saw what appeared to be live dinosaurs cavorting on the screen! Conan Doyle refused to answer questions about the film, other than to insinuate that what may at first appear to be impossible, may indeed be possible. His message to the magicians was that they should not dismiss phenomena they did not understand too quickly.

The footage was actually an excerpt from an upcoming movie based on Conan Doyle's novel *The Lost World* and was the first example of stop-motion photography of miniatures. Doyle had the last laugh on this one, as none of the magicians, including Houdini, were able to explain how dinosaurs had been captured on film. Famed science writer Arthur C. Clarke would later comment, "Any sufficiently advanced technology is indistinguishable from magic." The spiritualist Conan Doyle had clearly demonstrated this to the skeptical magicians.

As correctly portrayed in the television version, Conan Doyle believed that Houdini actually had psychic powers because he could not imagine how the magician's wondrous stunts could be performed by scientific means. This, in spite of knowing how science had done the apparently impossible by bringing his fictional dinosaurs to life. Too bad the factual episode of the dinosaur footage at the magicians' conference didn't make it into the History Channel's *Houdini.* As a fan of both Houdini and Sherlock Holmes, I was a little annoyed by the drifting from facts, but nevertheless was glad to see the program's spectacular ratings, which means that more people will be stimulated to explore the legacy of Houdini's magic, his escapes and his crusade against charlatans.

CHEMICAL WORRIES

KNOWING WHEN TO WORRY!

I don't think Einstein had chemical anxiety or the number of chemicals in our urine in mind when he famously stated, "Not everything that counts can be counted and not everything that can be counted counts." But I think the quote has great relevance given that scarcely a day goes by without some concerned group clamoring about our exposure to "untested" chemicals and lamenting the "fact" that we have become a nation of "unwitting guinea pigs."

Our exposure to chemicals is indeed extensive. Eat a bowl of chicken soup and hundreds of chemicals will flood your bloodstream. They include such delights as benzene, methanol, acetaldehyde and hydrogen sulfide, all of which are potentially "highly toxic." Of course they are not toxic in the dose found in the soup. But should you look for them in the urine, thanks to our sophisticated analytical techniques, you will find them. Nobody bothers to look, because these chemicals are not deemed important — after all they are "natural," and nobody has a political interest in banning chicken soup. But the story is different when it comes to synthetic compounds, especially those that have been deemed to be endocrine disruptors.

When I was alerted to a *New York Times* article featuring the headline "How Chemicals Affect Us," I was pretty sure the columnist was not about to discuss how antibiotics cure infections, how preservatives protect us from eating moldy food or how detergents clean our clothes. I knew I'd be reading a litany of warnings about toxins, poisons and endocrine disruptors. Unfortunately, that's what the term "chemical" has come to mean. To many, chemicals are the substances that insidiously invade our lives and shorten them.

The columnist was right about one thing. Chemicals do affect us! And they do so in every conceivable way! Take away oxygen and you die. Eat an improperly processed puffer fish, and the natural tetrodotoxin it harbors will kill you. If you have a headache, aspirin comes in handy. And, yes, chemicals that leach out from plastics can have hormonal effects. But "hormonal effect" is not synonymous with "hormonal disruption." An effect on a cell in the laboratory cannot be uncritically extrapolated to what may happen in a living body.

The problem is that the human body is very complex, and its interaction with the environment is virtually impossible to totally clarify. We are exposed to a vast array of compounds, and how they interact with each other and with the naturally occurring compounds in our body, defies analysis. One can take virtually any single compound, carry out animal studies with varying doses and find something that can be used to raise alarm. The bottom line is that nobody really knows, because the effects of trace chemicals cannot be teased out from the biological noise. This is especially the case for chemicals that mimic hormone activity. Is bisphenol A worse than soy? Do we stop drinking milk because it contains estrogens? Do we ban alfalfa sprouts because they contain coumestan, an estrogen mimic?

Critics who target one class of substances are unaware of the chemical complexity of life. Let's try an analogy. Suppose you're listening to a symphony orchestra and one string on a violin breaks. Do you think anyone would notice a difference in the sound? I doubt it. Similarly, removing one compound from the thousands and thousands to which we are exposed is unlikely to have a significant effect on life. Unlikely, but not impossible. Basically, both sides of the endocrine disruptor debate imply that they know more than they actually know, or indeed, what can be known.

A recent study by the Silent Spring Institute, a non-profit research organization, is a case in point. Researchers enlisted twenty people who volunteered to have the amount of bisphenol A and phthalates in their urine measured before and after a change in their diet. For three days, they agreed to avoid all canned and packaged products and to build their diet around fresh, organic food. And guess what. After three days, bisphenol A levels and phthalate levels in the subjects' urine decreased by roughly 65 and 55 percent respectively. Wow! Looks like you can decrease these "toxic" chemicals in your body dramatically after just three days by avoiding processed foods!

But wait a minute. It's so easy to play with numbers. Want to increase your chance of winning the lottery by 100 percent? Sounds good? Just buy two tickets! Statistically you've doubled your chance, but does it matter? Doubling a very small number still leaves you with a very small number. Similarly, what does a 65 percent decrease mean if it is a decrease from a number that was tiny in the first place? And the amounts of BPA and phthalates were tiny. Way, way less than any regulatory limits. So what is the big deal about such a decrease?

In fact, what the results actually show is that these chemicals are cleared quickly from the body. But fear of these chemicals is not cleared quite so quickly. The stress caused by the constant harangue takes a toll on health, even though it cannot be measured the same way that levels of the chemicals in question can be measured in the urine.

Don't get me wrong. I'm a big proponent of eating fresh, unprocessed foods, but more for the beneficial nutrients they contain than the "toxins" they eliminate. Remember, though, that chicken soup made with fresh vegetables and organic, free-range chicken can still deluge the urine with plenty of compounds that could be vilified the same way as BPA or phthalates, if only one cared to make the effort. You put parsnips in your soup, don't you? Well, they contain psoralens, compounds that are not destroyed by cooking and have carcinogenic potential. But I'm not worried about the psoralens. Or about storing my leftover soup in plastic containers. Why not? Because I look at numbers. And those numbers tell me that whatever "toxins" may be present are there at levels that way below what regulatory agencies find acceptable. I know how the scientists at Health Canada, the FDA and the EPA determine these levels. I know their qualifications and level of expertise. I also know the same for their critics. I know whom to trust.

OUR UNDERSTANDING OF ENDOCRINE DISRUPTORS MOVES AT A SNAIL'S PACE

The researchers were surprised to see that snails reared in some plastic water bottles produced almost twice as many offspring as their brethren raised in glass bottles. This wasn't some experiment by restaurateurs looking to add more snails to the menu. There wouldn't be much point, since these were New Zealand mud snails, less than half a centimeter in size, with not much meat on them. But the snails are pretty meaty when it comes to research about endocrine-disrupting chemicals.

These little creatures are fascinating in that they can reproduce asexually. The females are born with embryos in their reproductive system ready to develop into baby snails when stimulated by natural hormonal activity. But the development can also be prompted by exposing the snails to an external source of estrogen or estrogen-like compounds. And that means the snails can be used to detect the presence of chemicals with hormonal activity, or as they are popularly called, endocrine disruptors.

This area of research was set ablaze in the 1990s with some scientists claiming to have found a link between such compounds and a variety of health effects ranging from reproductive and behavioral problems to obesity and cancer. The most widely investigated hormone mimic since then has been bisphenol A (BPA), used to make polycarbonate plastics and epoxy resin linings in food cans. But even with twenty to thirty new research papers cranking out new data every month, there is much controversy about what effect, if any, exposure to BPA, or indeed to any other hormone-like compound, has on human health. Virtually every paper ends by calling for more research.

Back in 2009, Professor Jörg Oehlmann at Goethe University in Germany answered the call by assigning graduate student Martin Wagner a project on endocrine disruptors. One day, Wagner had a refreshing idea as he contemplated the task while sipping from a bottle of water. The possibility of bisphenol A leaching from refillable bottles made of polycarbonate had already caused a great deal of commotion, but Wagner wondered whether plastics that were not made with BPA might also release endocrine disruptors. Hence the experiments with snails.

Oehlmann and Wagner's research found that snails immersed in a culture medium in bottles made of polyethylene terephthalate (PET), the common plastic used in single-serve water bottles, experienced an estrogen effect, while those raised in glass bottles did not. This observation spurred further research using actual samples of bottled water and a more sensitive test based on the reaction of a specific type of yeast with estrogens. By and large, the results mirrored the snail findings, verifying the presence of estrogenic compounds. Of course the presence of such compounds cannot be equated with the presence of harm. Let's remember that we are awash in estrogen mimics, both natural and synthetic. Beans, lentils, sesame seeds, tea, peanuts, milk and various medications contain estrogenic compounds.

The interesting conundrum here is why bottled water should have any hormone-like activity at all. The snail experiment implies that chemicals are being leached from the plastic, but that's a puzzler since neither ethylene glycol, nor terephthalic acid, the two compounds used to make PET, exhibit hormone-like activity. The answer may lie in the complexity of plastics — one greater than is implied by the popular definition that these are materials composed of giant molecules (polymers) made by joining small molecules together in a chain. In actual fact, there are literally dozens of additives, ranging from antioxidants and polymerization catalysts to color modifiers and ultraviolet light absorbers, some of which may well have estrogen-like properties. Exactly what additives are used is hard to determine since plastic formulations are proprietary. The German researchers believe they may have identified di(ethylhexyl)fumarate as one of the compounds in bottled water that blocks estrogen receptors on cells, but its origin is a matter of mystery.

The complexity of the endocrine disruptor controversy is further illustrated by a lawsuit that was brought against two Texas companies by Eastman Chemical, manufacturer of Tritan, a hard, clear plastic that is marketed as BPA-free and estrogen activity-free. Scientists from CertiChem and PlastiPure had published a paper in which they reported that many plastics advertised as being free of any estrogen activity actually leached endocrine disruptors, especially when the plastics were exposed to heat or ultraviolet light.

CertiChem's testing method was questionable, Eastman argued in the lawsuit, also pointing out that the company was associated with PlastiPure, which sold plastics guaranteed to be free of such activity and was set to benefit from claims that competitors' products were actually contaminated with endocrine disruptors. At the trial, both sides trotted out an array of experts who flung accusations at each other. The "independent" tests presented by Eastman that showed no estrogen activity were tainted, PlastiPure's witnesses claimed, because the company had paid for the testing. Eastman acknowledged that it had paid for the study, but emphasized that it had no further involvement. More importantly, the study was done on animals, which Eastman pointed out was more meaningful than CertiChem's study using cultured cell lines.

After all was said and done, the jury found for Eastman. While this case was settled in the courtroom, endocrine disruptors remain on trial in the public arena with alarmist charges of great threat to our health being countered with comforting words and supporting data from regulatory agencies. At current exposure levels, Health Canada and the U.S. FDA maintain endocrine disruptors are of no great concern. Unfortunately, as with many such controversies, the consumer is left to struggle with the conflicting views.

At this point, the only conclusion I'm prepared to draw, and that with no great conviction, is that routine heating of foods in plastic is not advisable. As far as bottled water goes, concerns about the environment trump worries about estrogenic activity when it comes to avoidance. But if you want to raise New Zealand mud snails, it seems plastic water bottles are the way to go. Make sure you keep the creatures captive though, because in the wild they become an invasive species, squeezing out native snail populations that serve as food for fish.

NANOSILVER IS NOT A NANOPROBLEM

Living is a wasteful business. Just think of all the stuff we dispose of down our drains and toilets. Pharmaceuticals, oils, cosmetics, hair, condoms, glues, paints, nail polish removers, soap, urine, feces, food remnants, toilet paper, pesticides, dyes, cleaning agents, blood and vomit. And there are plenty of bacteria and viruses that go down as well. The original answer to this onslaught to nature was that the "solution to pollution is dilution." Basically that meant the mess would be diluted enough in natural water systems that it would not come back to bite us in the rear when it came to recycling water. And of course water has to be recycled. They just don't make the stuff anymore. But as the complexity of society increased, and as more and more waste was generated, novel technologies had to be introduced to deal with wastewater.

Numerous methods are used to deal with sewage, but they all involve some sort of holding tank to allow solids to settle to the bottom and oily substances to float to the surface, where they can be skimmed off. The water is then exposed to various microbes that are adept at decomposing organic waste. The sludge that remains is either buried or, in some cases, is used as fertilizer. Any remaining contaminants in the sludge, such as silver nanoparticles, can be a problem.

What are silver nanoparticles? Imagine taking your silver candlestick and putting it into some sort of grinder and grinding away until it has been reduced to tiny particles, so small that each one is less than one billionth of a millimeter in size. And on this scale, the "nano" scale, you'll be looking at a different world. Of course, one single nanoparticle is way too small to be seen, but the collage of nanoparticles of silver will no longer appear as a bright, shiny metal because the tiny particles now absorb and reflect light in a different fashion, giving them a yellow color. Long before anything was known about nanoparticles, artisans ground silver finely and added the particles to glass to give church windows a permanent yellow color.

As particle size decreases, the ratio of the surface area to volume increases dramatically. Just a pinch of nanoparticles has a gigantic combined surface area. And this is what plays a pivotal role in the antibacterial effect attributed to nanosilver and explains why it is incorporated into textiles. The goal is to kill bacteria that are responsible for producing body odors, which are the result of bacterial action mostly on the fats found in sweat. Enzymes called lipases produced by bacteria break down fats to yield small, odiferous molecules such as butyric, propionic and isovaleric acids.

Insoles of athletic shoes are also sometimes treated with nanosilver particles, as are the inner surfaces of some appliances such as refrigerators, washing machines and air conditioners. In all these cases, the intent is to make use of silver's antibacterial and antifungal properties, which are now well understood. It turns out that metallic silver itself is inactive, but in the presence of moisture and oxygen, a process known as oxidative dissolution occurs. This leads to the formation of positively charged silver ions, a result of silver atoms losing an electron to oxygen. It is these ions that interfere with bacterial metabolism.

But there is a degree of concern with nanosilver particles being used in a wide array of consumer items because when textiles are washed, or when appliances are discarded, some of the silver, and consequently silver ions, ends up in wastewater. The resulting antibacterial action can interfere with the work of microbes in the treatment plants. And if sludge from these plants is used as fertilizer, which is common because of its high phosphorus content, soil microbes, including nitrogen fixing bacteria can suffer, causing damage to agricultural land. It has been estimated that if everyone buys just one silver-particle treated pair of socks a year, the silver concentration in wastewater sludge can double. Of course this is not a realistic possibility.

There are other issues that arise with the commercialization of nanosilver, including some unsubstantiated hype. For example, some washing machines advertise that the billions of silver ions released in each wash cycle kill 99.99 percent of bacteria. But regular washing in hot water also kills over 99 percent of bacteria. Quibbling over a less than 1 percent difference in antibacterial effect is hardly good science. There are also questions about just how effective nanosilver in fabrics or shoes actually is in terms of preventing smells, and surveys have also shown that some products that say they contain silver don't, and some that don't make the claim do. And in many cases the silver is actually not nano, which would reduce its disinfectant efficacy because, as already mentioned, it is the extremely high surface area to volume ratio that makes nanosilver an effective antibacterial agent. And if the silver isn't effective enough, it can actually have an opposite effect to that desired. Exposure to sub-lethal doses of silver ions can improve bacterial survival rates and increase the chance of resistance.

While nanoparticles of silver are unlikely to pose any hazard to people, the same cannot be said for colloidal silver solutions that are promoted as a dietary supplement to stave off illness. Here we are talking about a much greater exposure to silver with the possibility of developing argyria, a permanent discoloration of the skin due to deposits of silver sulfide and silver selenide. When the colloidal silver particles hit the acidic environment in the stomach, they undergo oxidative dissolution and the resulting silver ions form complexes with glutathione, a naturally occurring antioxidant in the body. This complex is readily transported around the body by the bloodstream, and when exposed to ultraviolet light near the surface of the skin, undergoes a photo-reduction process whereby the silver ions are converted to metallic silver in the form of silver nanoparticles. These then react with sulfur and selenium compounds in the body to form the gray deposits that characterize argyria. And no silver polish will get rid of that.

PARAXYLENE PARADOX

Chinese cities don't often see demonstrations. But the proposed building of a new chemical plant in the city of Kunming did manage to drive thousands of people into the streets in protest. Kunming may be small by Chinese standards, but it has a population of six million, many of whom don't want a giant paraxylene plant in their backyard. They're worried about the release of toxic chemicals, which is understandable given that a recent analysis by the Health Effects Institute in Boston found that every year over a million people die prematurely in China due to air pollution. The fear is that the new chemical plant, on track to process 10 million tons of crude oil and produce half a million tons of paraxylene (PX for short) every year, will contribute to pollution. So, why is there a need to produce such vast quantities of a chemical that most people have never heard of? Because it is needed to make terephthalic acid, one of the main components of a substance you surely have heard of, namely, polyester!

Our life without polyester would be dramatically different. Bottles of all kinds, synthetic fibers, insulating materials and fiberglass are just some of the consumer items made from this synthetic polymer. Whenever you see the #1 recycling logo on a bottle, you're looking at polyethylene terephthalate (PET), perhaps the most widely used polyester. Coca-Cola alone uses about one hundred billion polyester bottles a year! That's about fourteen bottles for every person on Earth. Some are recycled into items such as insulation, outdoor furniture or easy to care for and comfortable microfiber fabrics. Recycling, though, cannot meet the demand for polyester, hence the need for paraxylene production.

The source for paraxylene, as it is for so many of the chemicals needed for modern life, is that mixture of hundreds of compounds formed by the slow decomposition of organic matter we call petroleum. Distillation of crude petroleum yields a mixture of liquids known as naphtha, which can be subjected to further catalytic reforming to produce benzene, toluene and a mixture of closely related compounds known as xylenes. This somewhat curious name derives from the Greek word for wood, "*xulon*," because the colorless, flammable, pleasant-smelling xylenes can be obtained from the distillation of wood in the absence of oxygen. The mixture of xylenes can be refined to yield paraxylene, which in turn is converted to terephthalic acid, the compound used to make polyethylene terepthalate. Xylenes themselves are often used as paint thinners, degreasers, cleaning agents and solvents for dyes, pigments and pesticides.

There are two fundamental issues here. First, how likely is the release of chemicals from a paraxylene plant, and second, if there is some release, what is the risk? Modern chemical facilities are built according to stringent specifications and feature many safeguards. The risk of releasing significant amounts of chemicals into the environment is small, but not zero. People, however, clamor for zero risk, a fundamentally flawed concept. Pipelines can leak, trucks, trains or ships with chemical cargoes can be involved in accidents. Earthquakes, tornadoes, hurricanes and tsunamis can have catastrophic consequences for chemical manufacture. Humans can make errors. But, as is always the case, it comes down to a proper risk-benefit analysis. Do we want to give up our cheap and shatterproof bottles, insulating materials, comfortable apparel, carpeting, upholstery, blankets, draperies, sheets, pillowcases, seatbelts, ropes, protective packaging for fruits and vegetables and synthetic artery replacements, all of which are made from polyester? I think not. But of course we want assurance that production is carried out in the safest possible fashion and that recycling is vigorously pursued. Still, no matter what safeguards are put into practice, the possibility of some sort of inadvertent chemical release cannot be discounted. That of course is the case not only for paraxylene, but also for the hundreds of other chemicals that are produced in plants around the world to feed the needs of modern society.

What then would be the consequence of paraxylene escaping into the environment? There is no simple answer because effects of course depend on the extent of exposure. Paraxylene, like other xylenes, is a neurotoxin which means it can produce headaches, dizziness and disorientation. This, however, requires continued exposure to concentrations in the air of well over 100 parts per million, something that has not been observed around paraxylene producing facilities. At far greater concentrations there can be liver and kidney damage.

Even at very low concentrations xylenes can irritate the eyes, nose or throat, something to which I can attest. I've had occasion to write on "whiteboards" with the special markers provided. The most common solvents used in these pens are toluene and the xylenes. They give me headaches and make me dizzy. I much prefer regular blackboards where your only concern is being covered in chalk dust. Whiteboard markers that use alcohol as the solvent are safer, but I still like chalk better.

While using a whiteboard marker can be annoying, abusing one can be life threatening. Deep inhalation, known as "huffing" can result in loss of inhibition, hallucinations, delusions and impaired judgment. Chronic abuse can cause irreversible damage to the heart, liver, kidneys and brain. Unfortunately such abuse is not infrequent, with surveys showing that about 20 percent of eighth graders admit to having tried inhalants as a method of producing a high at least once in their life. You can't get high from sniffing chalk. And unlike whiteboards, blackboards erase quite easily. But if you must use whiteboards, the best way to erase markings is with a microfiber cloth, made of, you guessed it, polyester.

Now you know why paraxylene is such a coveted chemical and also why there is concern in Kanming about its production. But the engineers who will design the plant have a pretty good motivation for getting things right. Chinese authorities have recently given courts the authority to hand down the death penalty for serious cases of pollution. That's motivation.

FLAME RETARDANTS – OUT OF THE FRYING PAN AND INTO THE FIRE?

Talk about a hot issue! Flame retardants in clothing, furniture and electrical equipment are supposed to protect us from fire, but according to some critics they create more problems than they solve. The argument is that exposure to these chemicals may be linked with a variety of health problems that outweigh any benefit that may be provided by preventing fires.

The cost of fire to society, both in dollars and lives, is huge, so fire prevention should of course be on the front burner. Back in 1975, seeing that every year about six thousand lives were being lost in the U.S. to fire, and damage totaled upward of $10 billion in current dollars, the California legislature decided that something had to be done. A law was introduced requiring certain furniture and some other products, such as children's clothing, to comply with flammability standards. The basic requirement was that the materials used had to withstand a small open flame for twelve seconds without catching fire. In order to meet that standard, manufacturers began adding chemicals belonging to a family known as polybrominated diphenyl ethers (PBDEs), which prevented quick ignition. The amounts were not trivial; some sofas contained as much as a kilo of the chemicals. Although the law applied only to California, manufacturers everywhere abided by it because of California's large customer base. It would have been too expensive to manufacture items differently for California and the rest of the world.

According to statistics provided by the National Fire Prevention Association in the U.S., the total number of fires dropped from almost three quarters of a million in 1977 to less than half a million in 2004. The number of deaths was cut in half and property damage costs dropped by about 40 percent. Just how big a role flame retardants played is hard to determine because smoke detectors were widely introduced and there were numerous changes in building codes. According to some studies, the protection provided by PBDEs was not a big factor.

The introduction of flame retardants was obviously well-intentioned but, as the proverb states, the road to hell is lined with good intentions. By 2004, flame retardants were feeling the heat not only from flames, but from scientific studies. Researchers linked PBDEs to problems in the development of the nervous system, disruption of the endocrine system, fertility issues, behavior problems and decreased birth weight. In animals PBDEs have been associated with decreased levels of circulating thyroid hormones and some human studies have linked blood levels of flame retardants with health problems. But such associations can never prove cause and effect: it may be that people with higher levels of flame retardants in their blood also were exposed to various other chemicals. No adverse effects have been noted in foam or electronic equipment recyclers or in carpet installers, despite these workers having higher blood levels of flame retardants than the average population.

Still PBDEs raised concerns because they are persistent chemicals and show up in indoor dust, commercially available foods and breast milk. The most troublesome flame retardant seemed to be pentabromodiphenyl ether, which was banned by California in 2003, prompting its manufacturer to voluntarily cease production. Other PBDEs are now being phased out, but because of slow turnover of products that were formulated with them, their long half-life and their potential to bioaccumulate, human exposure will continue for many years.

Since susceptibility to hormone-disrupting substances is greatest during infancy, there has been concern about exposure to flame retardants while children are restrained in car seats. Since the time spent in such seats may be considerable, there is the possibility of exposure from skin contact as well as from ingesting or inhaling contaminated dust emanating from the seats. Researchers in New Zealand investigated this issue by estimating exposure based on studies that measured flame retardant concentration in the air and dust inside cars. On comparing the amounts that could be expected to be inhaled or ingested to animal studies that showed a risk, they concluded that flame retardants in car seats were not likely to cause adverse health effects.

Gymnasts also make for an interesting example of continued exposure to flame retardants. Gyms are equipped with foam landing mats and pits filled with pieces of foam to protect gymnasts from injuries, particularly when practicing new moves. They commonly joke of "eating pit" after a less than perfect landing, which generally means being covered in dust after plunging into the pile of foam.

The most common foam is made of polyurethane, which is treated with flame retardants according to current fire safety regulations. It takes a beating in gyms, causing it to slowly crumble and release dust that is laden with flame retardants. This dust may be inhaled or ingested if hands are not properly washed before eating. While no health effects have been noted in gymnasts, concern has been raised because in a study of eleven gymnasts, blood levels of brominated flame retardants were found to be three times higher than in the general population. Nobody is recommending that gymnasts should curtail their activities, but gyms should be emphasizing the importance of washing hands after practice. A good idea for office workers as well, since office dust contains flame retardants, probably because of their presence in electronic equipment.

The world of flame retardants is set to undergo a significant change because on January 1, California eliminated the twelve-second exposure to an open flame test in favor of a requirement that upholstered furniture must not continue to smolder some forty-five minutes after a lit cigarette is placed on it, a seemingly more realistic scenario. Importantly, manufacturers can meet the requirement without the use of fire retardants by choosing appropriate fabrics and fillers. As expected, the flame retardant industry opposes this change vehemently, claiming that it will result in an erosion of safety. That's debatable, as is whether a decrease in exposure to a few compounds with possible hormonal effects will have an impact on our health in a world where exposure to thousands of both synthetic and natural hormone mimics is common. Hopefully with the elimination of brominated flame retardants we won't be going from the proverbial frying pan into the fire.

BLOWING IN THE WIND

David Copperfield performed many an illusion on his television specials with his hair blowing in the wind, tousled by an off-stage fan. I was reminded of that effect by an episode of *The Dr. Oz Show* in which the hot air so often generated by the host was amplified by a fan à la Copperfield. And Oz too was performing a sort of illusion if we go by the definition of the term as "something that deceives by a false perception or belief." In this case, Oz dumped a bunch of yellow feathers on a patch of synthetic turf adorned with some synthetic plants to demonstrate pesticide drift. The flurry of feathers was meant to illustrate how neighboring fields, as well as people who happen to be nearby, might be affected. A powerful visual skit to be sure, but a gross misrepresentation of the risks posed by pesticide drift.

The reason for the demo at this particular time was that, in Oz's words, "the Environmental Protection Agency is on the brink of approving a brand new toxic pesticide you don't know about." The reference was to Enlist Duo, a pesticide that at the time was already approved in Canada. It is actually a mixture of the weed killers glyphosate and 2,4-D (short for 2,4-dichlorophenoxyacetic acid) designed to be used on corn and soy grown from seeds genetically engineered to resist these herbicides. Fields can then be sprayed to kill weeds without harming the crops.

The need for the new combination was generated by the development of resistance to glyphosate by weeds in fields planted with crops genetically modified to tolerate this herbicide. Such resistance has nothing to do with genetic modification; it is a consequence of biology, since some members of a target species will have a natural resistance to a pesticide and will go on to reproduce and yield offspring that are also resistant. Eventually the whole population becomes resistant. This is the same problem we face today with bacteria developing resistance to antibiotics.

Oz got one thing right. Pesticides are toxic. That's exactly why they are used. And that is why there is extensive research about their effects, and strict regulation about their application. Remember that there are no "safe" or "dangerous" chemicals, just safe or dangerous ways to use them. As far as 2,4-D and glyphosate go, there is nothing new here since both of these have been widely used for years, although not in this specific combination. What is new is the development of crops resistant to 2,4-D, which will allow for its use to kill weeds in corn and soy fields, something that was not possible before. This has raised alarm among those who maintain that 2,4-D is dangerous and that its increased use is going to affect human health. Dr. Oz apparently is of this belief, and as the feathers were flying around the stage, he chimed in with how "2,4-D is a chemical that was used in Agent Orange, which the government banned during the Vietnam War."

2,4-D was indeed one of the components in the notorious Agent Orange used to defoliate trees in Vietnam. Tragically, Agent Orange was later found to be contaminated with tetrachlorodibenzodioxin (TCDD), a highly toxic chemical linked to birth defects and cancer. This dioxin, however, has nothing to do with 2,4-D: it was inadvertently formed during the production of 2,4,5-trichloroacetic acid, or 2,4,5-T, the other component in Agent Orange. That is why the production of 2,4,5-T, but not 2,4-D, was banned!

It is deceitful to imply that the new herbicide is dangerous because it contains a harmful compound that was used in Agent Orange. Not only does Enlist Duo not contain any TCDD, the form of 2,4-D it does contain is also different from what was used in Vietnam. Enlist Duo is formulated with 2,4-D choline, which is far less volatile than 2,4-D itself and has an even safer profile. While legitimate concerns can be raised about genetic modification, it is disingenuous to scare the public by linking the newly proposed herbicide to Agent Orange. It is also irresponsible to show videos of crops such as green peppers being sprayed, insinuating that Enlist Duo will be used on all sorts of crops, whereas it would only be suitable for Dow's genetically engineered corn and soy.

Now on to the issue of pesticide drift, which can happen in two ways. Tiny droplets of a sprayed pesticide can be carried by air currents, and the chemicals from the spray can also evaporate and spread as a vapor after being deposited on a field in their liquid form. These are realistic concerns, especially given that some schools are located in the vicinity of agricultural fields. But these are just the sort of concerns that are taken into account when a pesticide is approved. For example, one well-designed study concluded that a person standing about 40 meters from a sprayer would be exposed to about 10 microliters of spray, of which 9 microliters are just water. Calculations show that the amount of 2,4-D in the 1 microliter is well within safety limits, and of course spraying isn't continuous: it is done a few times a year. Consider also that 2,4-D choline, which is what is found in Enlist Duo, has far lower volatility and tendency to drift than 2,4-D itself, further improving its safety profile.

Another factor that is taken into account before a pesticide is registered for use is its mode of action. Glyphosate, for example, interferes with the synthesis of amino acids needed by plants to produce vital proteins. The pathway by which these amino acids are produced is not found in animals. Humans have no need for such biochemical synthesis because the amino acids we require are supplied by our diet. As far as 2,4-D goes, it mimics the action of a plant hormone and causes rapid growth of plants that cannot be sustained by available nutrients, and the plant withers and dies. Such plant hormones have no human equivalent. This of course does not mean that these substances cannot cause harm by some other mechanism, but nevertheless the fact that their mode of action is through processes not present in humans is reassuring.

While no pesticide can be regarded as risk free, the portrayal of Enlist Duo by Dr. Oz amounts to unscientific fear mongering. His final comment that "this subjects our entire nation to one massive experiment and I'm very concerned that we're at the beginning of a catastrophe that we don't have to subject ourselves to" totally ignores the massive number of experiments that have been carried out on pesticides before approval — ones based on a scientific rather than an emotional evaluation of the risk-versus-benefit ratio. True, when it comes to pesticides, there is no free lunch. But without the judicious use of such agrochemicals, producing that lunch for the close to ten billion people who by 2050 will be lining up for it becomes a challenge. What we need is rational discussion, not the spraying around of feathers and ill-informed rhetoric in a deception-laden stage act. If I want deception on the stage, I'll stick to watching David Copperfield.

WITH BISPHENOL A, THERE IS SMOKE, BUT IS THERE FIRE?

Back in the 1930s, Swiss chemist Pierre Castan was researching materials for denture repair and American Sylvan Greenlee was working on novel paints. Independently, the two hit upon epoxy resins, one of the most useful classes of chemicals ever developed. These resins find a wide array of uses in paints, flooring materials, dental sealants, printing inks, medical devices, electronic equipment, printed circuit boards and adhesives. Epoxy glues help hold our cars, airplanes, furniture, boats, skis and electronic equipment together. Without them, our world would literally become unglued.

But there is a sticky point with epoxies, namely that they are formulated with the infamous chemical bisphenol A (BPA). This has prompted a search for a replacement, particularly in canned foods, where epoxy resin is used as a coating to separate the metal from the contents. Such a barrier is critical because dissolved metal can catalyze decomposition and impart an undesirable taste to food. Furthermore, as the metal dissolves, pinholes can form in the can, opening an entry point for bacteria and undermining the whole canning process. Epoxy resins are ideal for the protective layer because they form an airtight seal, stand up to the heat and pressure of sterilization, do not chip even if the can is dinged and do not alter the taste or smell of food.

Epoxy resins are polymers, meaning they are composed of long chains of individual molecules, or monomers, joined in a chainlike fashion. One of these monomers is bisphenol A, notorious because of its estrogen-like properties. Depending on to whom one listens, bisphenol A is either responsible for a stunning array of conditions, ranging from cancer and heart disease to obesity, or is an innocent bystander, falsely accused. Because of the publicity that BPA has received, numerous experts have weighed in, joined by a plethora of bloggers — many of whom are ignorant of the difference between hazard and risk and have not the slightest familiarity with the basic principles of chemistry, biology and toxicology. As usual, the public is left stressed and confused.

Besides epoxy resins, bisphenol A is used to make thermal paper for cash register receipts, as well as polycarbonate plastics for sporting equipment, cookware, automobile parts and water coolers. We touch cash receipts, eat canned foods and are exposed to dust that contains BPA as a result of abrasive contact with epoxy-based flooring, glues, paints and electronic equipment. Look for it and you will find it almost anywhere, except in blood, where it would matter the most. A 2014 Swedish study confirms earlier ones that found no detectible levels in the bloodstream of people and proposes that previous studies that did detect the chemical used contaminated samples. Indeed, BPA and its breakdown products are known to be quickly eliminated in the urine.

We are then left with the notion that levels in the urine are reflective of exposure, but not of biological activity, since that would require presence in the blood. Mennonite women who shun consumer products, don't eat canned foods and don't travel in cars do indeed have lower levels of BPA in their urine. However, they don't have a longer life expectancy and do not have a different disease pattern from the general population. Even if they did, it could not be ascribed to BPA because of genetic differences and totally different lifestyles. Nevertheless, many studies are based on measurement of BPA in the urine, and given that average the North American has detectible levels, and that there is no shortage of disease, a little data dredging can link

BPA to almost any condition one selects. Such associations can never prove cause and effect but they can create worry.

There are also animal and laboratory studies galore about BPA. But what do we make of them? You can take mice, implant human prostate cells, feed them BPA, treat them with estrogen to mimic the natural rise in aging men, and you can show an increase in prostate cancer. Whether the doses used reflect human exposure is questionable, as is the reliability of implanted cells in young mice being a model for what happens in an adult male.

Another rodent BPA experiment led to headlines such as "Is There a Link Between Migraines and Plastic," accompanied by pictures of plastic water bottles that are made of polyester and have nothing to do with BPA. In this study a tube was surgically implanted into the brains of female rats, apparently a procedure that creates migraine symptoms. I don't doubt it. When these animals were subsequently injected with BPA, they showed less movement, avoided loud noises and strong light, exhibited signs of tenderness to the head and were more easily frightened. I suspect anyone with a tube implanted in their head is likely to show signs of tenderness and fear. Such research gets you a publication, a few headlines, but does little to advance our knowledge of bisphenol A.

There's no doubt that when it comes to BPA there is smoke, but the big question is whether there is fire. The definitive experiment of course cannot be done since we can't use humans as guinea pigs and feed them chemicals of interest. So it comes down to relying on rodents, laboratory experiments and human epidemiological data to come to what amounts to a hopefully reliable educated guess. This is exactly what the European Food Safety Authority (EFSA) has just done.

After reviewing a wealth of studies, EFSA established a

tolerable daily intake (TDI) of 5 micrograms/kg body weight/day, ten times less than the previous TDI. A newer, more reliable method of estimating risk was used based on actual amounts of BPA known to affect the kidneys, liver or mammary glands of animals. The highest estimate for combined oral and non-oral exposure to BPA in people was found to be 30 to 50 percent less than the new, lower TDI. EFSA therefore concluded that exposure to BPA is too low to be of concern, the same conclusion arrived at by Health Canada and the FDA.

Still it would be desirable to replace epoxy resins in canned foods with substances that don't have the Damocles sword of "endocrine disruptor" hanging over their head. Some other plastics and vegetable resins are available for foods of low acidity like beans, yet so far none can match the efficacy and versatility of epoxy resins. But some ingenious chemist will undoubtedly solve the problem.

CHEMISTRY HERE AND THERE

MUSIC CHARMS BEASTS

Heart transplants are sometimes performed on rodents with the aim of testing anti-rejection drugs. But that's not what researchers at Teikyo University in Japan had in mind when they performed the operation on a group of male mice. They were interested in studying how the animals responded to different types of music piped into their "recovery rooms." This is not as outlandish as it may sound. Music has long been thought to have therapeutic properties.

The book of Samuel in the Bible tells us "whenever the evil spirit from God came to Saul, David would take the harp and play it with his hand; and Saul would be refreshed and be well, and the evil spirit would depart from him." The evil spirit was likely depression, and modern studies have corroborated the beneficial effect of music on levels of cortisol, the stress hormone. Undoubtedly undergoing a heart transplant is a stressful situation. Indeed, in humans studies have shown that patients who listened to music during and after open heart surgery required shorter intubation times. Such studies raise the question of whether different types of music lead to different outcomes, and that is precisely what the Teikyo researchers aimed to find out.

The study was carried out in a proper scientific fashion, with a control group compared to mice exposed to Verdi's *La Traviata*, a selection of Mozart sonatas, or songs by the Irish singer Enya. I would have liked to see another set of mice forced to listen to some loud rock, like the eardrum-bursting sounds that are blasted at spectators at hockey games the instant there is a stop in play. But that would probably have been too cruel. In any case, the results of the experiment were interesting. Mice that listened to Verdi or Mozart lived an average of twenty days longer than the animals that suffered in silence or the ones exposed to a single frequency tone. For some reason, the immune system of these animals was much more likely to reject the foreign tissue. Enya's songs were not much of an improvement over no music. It's hard to know what to make of such a study, but the mice may find some frequencies irritating and others more pleasing. Much like people's reaction to Red Hot Chili Peppers or Céline Dion.

Manufacturers of crystal singing bowls claim that people, like mice, also respond to specific notes and that "healing frequencies" can be generated by circling the rim of the bowl with a suede-covered mallet to produce an enchanting sound that eliminates the "disharmonious conditions" that cause disease. I can't get in tune with that. When it comes to health effects, I think people are far more likely to respond to music based on what they like rather than to specific frequencies. I know I would enjoy treatment with Andrew Lloyd Webber's "The Music of the Night" (especially if performed by Michael Crawford) a lot more than being exposed to the sounds of Limp Bizkit.

It seems that in my music preferences I may have something in common with egg-laying hens. British farmer Steve Ledsham was surprised when his chickens started laying eight eggs a week instead of the usual four. What was different? he wondered. The increased production seemed to coincide with the building of a new barn, suggesting it might have had to do with the music that was being played to entertain the workers. Ledsham now plays Webber's music all the time, and as he says, his farm now "is overrun with eggs." Soothing music, he feels, relaxes the birds and the calming effect increases egg production.

It isn't only chickens that perform better with music. It seems that cows produce more milk when they listen to calming music. And that isn't just hearsay. Researchers at the University of Leicester in the U.K. exposed herds of Friesian cattle to different types of music for twelve hours a day over nine weeks. On days when slow music was played, milk production increased by some 3 percent. Beethoven's *Pastoral Symphony* and Simon & Garfunkel's *Bridge Over Troubled Water* were big hits in the milking shed. On the other hand the cows did not enjoy "The Size of a Cow" by the Wonder Stuff. The music is pretty objectionable, and the cows probably did not think much of the simplistic lyrics either.

"A Moo Down Milk Lane" has no lyrics, but this original composition by Tzu-Deng Jerry D was judged to be the winning entry in a contest run by the British Columbia Dairy Association. The challenge was to come up with music that best increased milk yield, and apparently the cows really enjoyed Jerry D's dulcet tones. I wonder how the food that the cows chomp on would respond to this little composition. Yes, believe it or not, plant growth may also be affected by music. Back in 1973, Dorothy Retallack published a book called *The Sound of Music and Plants* in which she described her experiments that involved exposing plants to different types of music. "Easy listening" sounds actually made the plants lean toward the speaker, as if hungering for more. Rock music, on the other hand, frightened the plants, stunted their growth and caused them to seek refuge by leaning away from the speaker. The plants didn't care for country music one way or another, but interestingly, they did have a preference in terms of instruments. Strings, particularly the sitar, were favored over percussion instruments.

How can such a response be explained? Consider that, as far as we are concerned, sound is the brain's interpretation of the vibration of our ear drums in response to variations in air pressure. It is not inconceivable that such changes in pressure can have an effect on the movement of plant cells, resulting in changes in growth. This is more theory than hard science, but some vintners are convinced enough to have placed speakers in their vineyards, exposing the vines to the soothing sounds of Mozart and Vivaldi. I wonder what some of Justin Bieber's warblings would do. Maybe keep birds and insects away. There's an experiment waiting to be done.

TESTING FORMULA ONE

I'm not a huge fan of automobile racing, but I do admit to catching a bit of the fever when the Formula One cars roll into Montreal. There is something captivating about these machines capable of attaining speeds well over 180 miles per hour as they push technology, engineering and driving skills to the limit. This is not a cheap sport. The budget for a Formula One team can run upward of $120 million a year! Just the tires cost a couple of thousand for a set, and they only last for half a race. Of course these are not ordinary tires. They are made from a variety of specialized rubber compounds that can provide tremendous gripping power, the choice of specific tire being determined by weather conditions. The tires grip better as temperature increases and sometimes pre-heating is required.

There seems to be some confusion about what "Formula One" means. It does not refer to the fuel that is used. But the composition of the fuel is part of the formula, which actually refers to the set of regulations that define every aspect of the car as well as how the race is to be run. Perhaps surprisingly, the fuel used is regular gasoline. According to Formula One stipulation, it cannot contain any component that is not available in commercial gasoline, but the exact composition can vary subject to strictly defined limits.

As in any sport, cheating is always a possibility, and Formula One automobile racing is no exception. In this case, though, it is not only the drivers who have to provide samples to be tested for doping, but their cars as well in the form of gasoline. Contrary to common belief, gasoline is not a single chemical entity, rather it is a complex mixture of compounds derived mostly from petroleum, with smaller amounts of biofuels such as ethanol, produced through the fermentation of sugars. There are also various oxygen-containing additives designed to boost performance and detergents such as alkylamines and alkyl phosphates to protect the engine from the buildup of sludge.

The first stage of gasoline production begins with distillation of petroleum to capture compounds within a boiling point range that encompasses those having from four to twelve carbon atoms per molecule. Alternatively, higher boiling fractions can be subjected to catalytic cracking, causing larger molecules to break down to smaller ones typically found in gasoline.

In an internal combustion engine, organic compounds burn to yield carbon dioxide and water vapor, the gases that create the pressure needed to drive the pistons. In reality, combustion is never complete, and sometimes the unburned hydrocarbons can autoignite and cause the engine to "knock," resulting in reduced efficiency. One way to counter this problem is through the addition of lead compounds, a practice that has been phased out because of the metal's toxicity.

An alternate approach is to reformulate the gasoline by using specific catalysts to rearrange the atoms in some of the molecules to form compounds that burn more efficiently. Benzene, for example, belongs to a family of compounds known as aromatics and burns very well, but it is carcinogenic and the amount allowed in gasoline is limited. There is also the possibility of adding compounds from other sources to improve combustion. Ethanol, methanol, methyl-t-butyl ether (MTBE) and ethyl-t-butyl ether (ETBE) are some examples that enhance the efficiency of combustion because of their oxygen content. Ethanol has the added benefit of being made by fermentation of sugars from renewable resources such as corn. Obviously because of the number of compounds possibly present, the variety of blends of gasoline is practically infinite.

In the case of F1 fuel, the amount of aromatics, olefins (molecules with carbon-carbon double bonds) and compounds containing oxygen are all regulated. There is even a stipulation that a minimum 5.75 percent of the components must come from a biological source — in other words, not petroleum. Even though the characteristics of the fuel must conform to stringent guidelines, there is still enough maneuvering in exact composition to make a significant difference when it comes to racing.

Before a race, each team must provide a sample of the fuel to be used to the sport's governing body, the International Sports Car Federation (FIA), for analysis by an instrumental technique known as gas chromatography. Another sample taken at the event also has to be submitted. Each sample is injected into the gas chromatograph with a syringe and is immediately heated and vaporized. An inert carrier gas, usually helium, then pushes the vapors into a column filled with a packing material to which components of the mixture bind to different extents, meaning that they emerge from the column at different times. These retention times are characteristic of each component. The exiting gases are electronically detected and translated into a series of peaks on a chart paper, with the number of peaks representing the number of compounds detected, and the areas underneath the peaks being proportional to the relative amounts of each component. This output is then compared with one generated by a standard sample of a reference fuel, and if the variation is greater than specified by the rules, the fuel is deemed incompliant, and the car may be disqualified.

Chemical manipulation of the drivers can of course also make a difference. Driving one of these machines that pushes technology to the limit is physically and mentally demanding. FIA adheres to the World Anti-Doping Agency's protocols, and drivers are often tested during race weekends. But they may also be subjected to unscheduled tests outside of competitions to ensure they are not using drugs such as steroids to strengthen their muscles, or other performance enhancers. Transgressions are rare.

In case you think F1 racing is a total waste of money, well, it isn't totally. The technology developed has resulted in some useful spin-offs ranging from magnetic filters to remove rust from home heating systems to slip-resistant footwear. The telemetry systems that monitor 150,000 measurements a second from more than 200 sensors on an F1 car have been adapted to telemetry systems that help researchers monitor a variety of body functions in subjects taking part in clinical trials. Sometimes Formula One is a formula for scientific advancement. Now, if only they could do something about the noise . . .

SMARTENING UP ABOUT SMART METERS

Cell phones, microwave ovens, wi-fi, smart meters. What do they have in common? They all emit radiation in the radiofrequency range. And they all radiate controversy. Given that these devices are set to become as commonplace as lightbulbs, it is understandable that questions arise about their possible health effects. There are all sorts of allegations that exposure can trigger ailments ranging from headaches to cancer. Allegations, however, do not amount to science. And there is a lot of science to be considered.

Let's start with the fact that an alternating current flowing through a wire generates an electromagnetic field around it. This field can be thought of as being made up of discrete bundles of energy called photons that are created as the electrons in the wire flow first in one direction then in the other. Photons spread out from the wire, their energy depending on the frequency with which the current changes direction. The number of photons emitted, referred to as the "intensity" or "power" of the radiation, depends on the voltage, the current and the efficiency of the circuit to act as an antenna.

In ordinary household circuits, the direction of the current changes sixty times a second. In other words, it has a frequency of 60 Hz, the unit being named after Heinrich Rudolf Hertz, the first scientist to conclusively prove the existence of electromagnetic waves. The photons emitted by such a circuit travel through space and have the capacity to induce a 60Hz current in any conducting material they encounter. Essentially, we have a transmitter and a receiver. If special circuitry is used to produce current in the range of 10,000 (10KHz) to 300 billion Hz (300 GHz), the photons emitted are said to be in the radiofrequency region of the electromagnetic spectrum. That's because with appropriate modulation at the transmitter (amplitude modulation [AM] or frequency modulation [FM]) these photons can induce a current in an antenna that can be converted into sounds or images.

But what happens when photons in this energy range interact with living tissue, such as our bodies? The greatest concern would be the breaking of bonds between atoms in molecules. Disrupting the molecular framework of proteins, fats and particularly nucleic acids can lead to all sorts of problems, including cancer. However, photons associated with radiofrequencies do not have enough energy to do this, no matter what their intensity.

An analogy may be in order. Consider a weather vane sitting on a roof. It is mounted on a sturdy metal rod but of course can spin. You decide you want to knock it off the roof, but all you have are tennis balls. You start throwing the balls, but even if you hit the support, nothing happens. You just can't impart enough energy to the ball to have it break a metal rod. And it doesn't matter if you gather all your friends, and they all throw balls at the same time. You may have increased the intensity of your efforts, but it doesn't matter, because no ball has enough energy. Of course if you had a cannon, you could knock down the target with one shot. That's why high-energy photons such as generated by very high-frequency currents, as in x-rays, are dangerous. They can break chemical bonds! While you are not going to damage the weather vane with the tennis balls, you can surely make it spin, and the friction generated will heat up the base, the extent depending on how many balls are thrown.

Now, back to our photons. In the radiofrequency region, no photon has enough energy to break chemical bonds, but they can make molecules move around, generating heat. The more photons released, the greater the heating effect. This is exactly how microwave ovens work. They operate at radiofrequencies, but at a very high intensity or power level, meaning they bombard the food with lots of photons, causing the food to heat up. You certainly wouldn't want to crawl into a working microwave oven and close the door behind you. Similarly, you wouldn't want to stand right next to a high power radio transmitting antenna, such as those used by radio or TV stations, because you could get burned very badly. But the number of photons encountered drops very quickly with distance as they spread out in all directions, so that even standing a few meters from the base of such an antenna would not cause any sensation of heat. Just think of how quickly the heat released by a lightbulb drops off with distance.

The smart meters that are being installed by electrical utilities monitor the use of electricity and relay the information via a built-in radio transmitter. But the radiation to which people are exposed from these meters quickly drops off with distance, as with the lightbulb, and is way below established safety limits. Furthermore, the smart meters only transmit for a few milliseconds at a time for a grand total of a few minutes a day! Cordless phones, cell phones, routers, baby monitors, video game controls and especially operating microwave ovens expose us to similar radiation, usually at far higher levels. Smart meters are responsible for a very small drop in the radiofrequency photon bucket.

It must be pointed out, though, that safety standards are essentially based on the heating of tissues. But what about the possibility of "non-thermal" effects? What if radiofrequency photons cause damage by some other mysterious mechanism? Over the last thirty years, more than 25,000 peer-reviewed papers have been published on electromagnetic fields and health, many devoted to non-thermal effects. Health agencies do not find present evidence persuasive of a hazard at ordinary exposure levels, and given the extent of research that has been carried out, it is unlikely that one will be identified in the future.

Although an overwhelming number of studies on cell phones and brain cancer have shown no effect, admittedly some have suggested a barely detectable link. Despite the weak evidence, the International Agency for Research on Cancer has classified electromagnetic fields associated with radiofrequencies as "possibly carcinogenic," indicating a level of suspicion without any implication that the fields actually cause cancer. This notion pertains to cell phone use and has nothing to do with the far weaker fields associated with wi-fi and smart meters. I would have no issue with a smart meter in my house.

What then about those consumers who claim they have developed symptoms after smart meters were installed? I think it is appropriate to consider John Milton's poetic view of the power of imagination: "The mind is its own place, and in itself can make a heaven of hell and a hell of heaven."

BAMBOOZLING LABELS

Was I bamboozled? No. The truth is that I had bought a "bamboo" t-shirt on a Caribbean cruise not because of environmental consciousness but because I liked its silky feel. However, the Federal Trade Commission in the U.S. believes that many consumers purchase clothing items that are advertised and labeled as bamboo because they feel they are buying a greener product based on bamboo's quick growth and lack of requirement for pesticides. So why did the Federal Trade Commission send a warning letter to a number of retailers who were selling bamboo products? Because the items labeled and advertised as bamboo were actually made of rayon!

There is nothing at all wrong with rayon fiber, but there is something wrong with misleading consumers by implying that the "bamboo" item they purchased is woven from fibers stripped from the bamboo plant. This is not the case. The production of rayon requires extensive processing. A variety of plants, including bamboo, can be used as a source material to create rayon, but the final properties of the fabric do not depend on what plant was used. And no matter what the raw material, the manufacture of rayon involves the emission of air pollutants and the use of a variety of chemicals that are not exactly environmentally friendly. Still, rayon is an excellent fiber and has played a significant role in the development of textiles.

Rayon is best described as regenerated cellulose. Cellulose is the most abundant organic chemical in the world, being the structural component of the cell wall of green plants. In terms of molecular structure, it is composed of anywhere from several hundred to over ten thousand glucose molecules linked together in a linear fashion. Cellulosic fibers from the flax plant were woven into linen as early as 8000 BC, and cotton, which has a cellulose content of over 90 percent was being grown, spun and woven into cloth by 3000 BC. Around the same time, the Chinese discovered that the cocoon formed by silkworm larvae could be unraveled and used to make very fine fabrics. Wool, matted into felt, also had its origins around this time. But all of these fibers had a downside. Cotton and linen wrinkled from wear and washings, silk required delicate handling, wool shrank, was itchy and served up a meal for moths. Still, until the nineteenth century clothing manufacturers had to put up with such problems because there were no alternatives to the natural fibers.

Until the late 1800s, if you wanted the luxury of silk you had better know where to find some mulberry trees infested with silkworm pupae that had wrapped themselves in a cocoon of raw silk they exuded from their salivary glands. Boiling the cocoon killed the pupae and prevented the secretion of enzymes that would normally break down the silk in order to allow the emergence of the adult in the form of the silk moth. These enzymes degrade the silk fiber from hundreds of meters in length to shorter segments, ruining it for fabric use. Some 5,000 cocoons are needed to make a kilo of silk which explains why the fiber is so pricey. Although historically various attempts had been made to bypass the silkworm and convert mulberry leaves into silk, they all failed. Finally, in 1889 at the Paris Exhibition, visitors got their first glimpse of a silk-like fabric that that was washable like cotton but had the luster and delicacy of the authentic material.

Count Hilaire von Chardonnet was the inventor responsible for the marvel that was to become the first marketable synthetic fiber. Silkworms or mulberry leaves were not involved, although the Count's interest had been spurred by his work with Louis Pasteur on silkworm diseases. He got the idea for his artificial silk by noting that collodion, the gelatinous substance formed by dissolving nitrated cellulose in a mixture of alcohol and ether, could be extruded into a fine thread. Chardonnet's contribution was weaving the threads into fabric. Unfortunately, Chardonnet silk had one major drawback. Like all nitrocellulose products, it was highly flammable. Workmen at the textile plants where the fabric was manufactured took to calling it "mother-in-law silk."

Around the same time, across the Channel in England, Charles Cross, Edward Bevan and Clayton Beadle were working on improving the manufacture of paper and cotton thread. The general process began by treating the cellulose-containing material with sodium hydroxide (lye) to extract the cellulose. Could other chemicals be added to produce a better product, they wondered? Yes, as became apparent in 1892, when alkali cellulose was treated with carbon disulfide to form a bright orange grainy substance that formed a viscous solution when dissolved in water. Wasn't of much use for making paper, but when this "viscose" was passed through tiny holes into an acid bath, it yielded a fiber that was still cellulose but in which the long chains of glucose molecules had been broken into shorter ones. The result was a fiber that until 1924 was commonly referred to as "artificial silk." That's when the DuPont Company began to produce this regenerated cellulose on a large scale, christening it "rayon," because of the fabric's lustrous, almost metallic sheen when the rays of the sun fell upon it. Textile manufacturers and their customers were thrilled because rayon was half the price of raw silk.

Today, rayon is widely manufactured for clothing, sheets, blankets and upholstery because it is smooth, cool and comfortable. And that is precisely why I bought my t-shirt. In fact I've ordered a couple more since. And I was gratified to find that the ones I bought from bamboo product retailer Cariloha would not have raised the FTC's ire. They were clearly labeled as "made of 70 percent viscose from bamboo and 30 percent organic cotton." No bamboozling here.

CAPTURING CARBON DIOXIDE

Carbon dioxide: is it the pollutant responsible for evils ranging from rising sea levels and terrifying storms to greater insect infestations, or is it a raw material that can reduce our reliance on petroleum for energy and chemical manufacture? The answer is both!

It is clear that we can no longer continue to spew carbon dioxide recklessly into the atmosphere. It is also clear that eventually we will run out of fossil fuels. So, how about killing two birds with one stone? (Only figuratively of course.) The task amounts to capturing the carbon dioxide formed in combustion processes before it is released into the atmosphere and then finding a way to convert it into useful compounds. After all, carbon is the basic building block of all organic compounds, be they in gasoline, pharmaceuticals or plastics, and carbon is of course one of the components of carbon dioxide.

There certainly is precedence for using carbon dioxide as a source of organic compounds. Just consider photosynthesis, the reaction that makes all life on earth possible by allowing carbon dioxide to react with water under the influence of the catalyst chlorophyll to yield oxygen and glucose. Plants then use the carbon framework of glucose for the synthesis of their myriad chemicals constituents. The complex reactions involved in photosynthesis have defied efforts at commercial replication, but a host of reactions have been developed that use carbon dioxide as a raw material for the synthesis of various organic compounds.

Research is complicated by the stability of carbon dioxide. Basically, this molecule does not easily engage in chemical reactions, as is of course witnessed by its buildup in the atmosphere. In chemistry, the usual way to coax unreactive molecules into activity is through the introduction of a catalyst, a substance that speeds up a reaction without undergoing a change itself. Chemists have long dreamed of finding the right catalyst for reactions that would allow abundant carbon dioxide to be used as a feedstock instead of non-renewable crude oil. That dream is now on the verge of becoming a reality.

Technically speaking, the capture of carbon dioxide from combustion effluent is the smaller part of the problem. Several technologies exist, with bubbling the carbon dioxide through the solution of an amine being a prime example. One reaction that carbon dioxide does engage in quite easily is with water to form carbonic acid. This doesn't happen to a great extent, which is why carbonated water is only slightly acidic. But amines are bases that react with acids, and as the carbonic acid concentration drops, more carbon dioxide reacts with water to replace the acids that have been mopped up by the amine. With enough amine in solution, eventually all the carbon dioxide is absorbed. The beauty of this process is that heating the solution liberates the carbon dioxide, leaving the amine solution available to be recycled. Obviously there is an energy cost to this process, about 20 percent of the total power output. Unfortunately, when it comes to reducing carbon emissions there is no free lunch.

What then is to be done with the massive amounts of carbon dioxide that can be captured by such processes? Some can be sold to soft drink manufacturers, but that is not a significant amount. For now, the plan is to inject the gas into geological formations several kilometers underground for safe storage. That doesn't seem to be the best way to deal with a potentially valuable material. Of course the key word here is "potentially," because while many research groups have found ways to convert carbon dioxide into useful organic compounds, the challenge remains to make these reactions practical on a large scale.

Some efforts are tantalizingly close. Bayer, the company whose name is mostly associated with aspirin, is close to finding a treatment for the headache that carbon dioxide has been causing researchers. After testing over 200 catalysts, its chemists came up with one that allows carbon dioxide to react with compounds in the epoxide family to yield polyether polycarbonate polyols (PPPs). Some may find these unpronounceable, but they are very functional. When reacted with isocyanates they form polyurethanes, one of the most useful classes of plastics. You'll find them as foams in furniture, mattresses and insulating materials, as well as in ski boots and the soles of running shoes.

Bayer has partnered with a power plant to obtain the carbon dioxide collected from its amine scrubber, which will be used to make polyurethane foam for mattresses. While the epoxides and isocyanates still have to be sourced from petroleum resources, the use of carbon dioxide results in an overall environmental benefit. The same can be said for polypropylene carbonate (PPC), a biodegradable plastic that can be made from epoxides and carbon dioxide for use in packing materials, linings for food cans and as a softener for more brittle plastics like polylactic acid.

And then there is the alluring possibility of turning carbon dioxide into ethylene glycol, a compound that has great commercial application as antifreeze and is also one of the components used to formulate polyester for water and soft drink bottles. Liquid Light, a New Jersey Company, has found a catalyst that will convert a solution of carbon dioxide to ethylene glycol as an electric current passes through it. Another possibility is to use a special molybdenum catalyst to couple ethylene with carbon dioxide to form acrylic acid, widely used to make adhesives, synthetic fibers, water treatment chemicals, Plexiglas and absorbents in diapers.

Using carbon dioxide to make chemicals does have the double-barrel effect of reducing greenhouse gas emissions while concurrently saving petroleum. However, the amount of petroleum used for chemical manufacture is a small fraction of that used for fuel. But waste carbon dioxide can even play a role in fuel production.

Algae are sensational little factories for oil production. Put them in a pond, supply some nutrients, pump in carbon dioxide and they'll provide a spectacular harvest. Then place them in a closed reactor with some sugar from food industry waste and they will plump up with oils that can be burned as fuel or converted into biodiesel. And finally, carbon dioxide can be captured and converted into sodium bicarbonate, which is just what is needed to treat the indigestion caused by comments from people who maintain that carbon dioxide emissions are not a problem.

CAT PHEROMONES

Once upon a time we had a cat. I think he was sort of smart because as soon as he got a glimpse of the cage we used to transport him in, he knew he was going to the vet. He didn't like that one bit and became nervous and agitated. Had a facial pheromone product been available back then, I surely would have sprayed the cage with it. Obviously this notion calls for a little background on cat chemistry.

First, a definition. Pheromones are chemicals that are produced by animals to affect the behavior of other members of the same species. Scout ants, for example, go out searching for food, and should a supply be found, they lay down a track of scent that leads other ants to the meal. The female silk moth wafts a chemical called bombykol into the air to attract a mate. When bees find a good source of nectar, they anoint the flower with a mix of citral and geraniol as signal to other bees to "come and get it." Male dogs know when a female is in heat because of the scent of para-hydroxymethylbenzoate is in the air. And then we come to cats.

If you think a cat is expressing its love when it rubs its cheeks on your leg, think again. You are just being marked as safe territory. When a cat feels comfortable, it deposits a chemical signal that allows it to easily relocate its safe haven, be it a comfy couch or be it you. A cage in which the animal is being transported is not likely to fall into this category, but what if it were anointed with a version of the facial pheromone? Could it possibly have a calming effect? Not an unreasonable notion. And possibly marketable.

Swabbing cat's faces to collect significant amounts of secretions for commercial use is not a viable option. But collecting enough for chemical analysis is. The facial secretions turn out to be a mix of fatty acids that include oleic acid, azelaic acid, pimelic acid, palmitic acid, butyric acid, caproic acid, 5-aminovaleric acid, para-hydroxyphenylacetic acid along with a dose of trimethylamine. Some of these have a pretty nasty smell if the concentration is high enough, with butyric acid conjuring up the smell of rancid butter and caproic acid broadcasting the aroma of a wet goat. Mercifully cat facial pheromones do not contain enough of these compounds to be detected by humans.

In terms of chemistry, all the components of a cat's facial secretions are pretty simple and are easily synthesized. These compounds have various applications in industry and are commercially available from chemical companies. Some inventive marketers have recognized this and have concocted proprietary mixtures that they claim have a calming effect on felines when applied to their surroundings. These mixtures also contain extracts of valerian, a plant known to be especially appealing to cats, with the idea that the extract will attract the animals to the area that has been sprayed with the "calming" mix of fatty acids. The specific valerian compound is actinidine, similar to substances found in catnip. Why such compounds should attract cats isn't clear, but it may be because they mimic some components of cat urine that is used to signal readiness for sexual activity.

Cats spray with urine not only as a sign of sexual availability, but also when they perceive a threat to their territory, such as when other cats enter the picture. They may also be disturbed by the smell of new carpets or furniture or changes in their diet. Can urine spraying caused by such stresses be reduced with the scent of the anti-anxiety facial pheromones? Marketers of such products say yes. Evidence, however, is less than conclusive. For example, researchers at the University of Edinburgh Hospital for Small Animals organized a randomized, double-blind, placebo-controlled trial to study nine cats suffering from feline idiopathic cystitis, a condition associated with non-sexual urine spraying. The cats had their environment treated daily either with feline facial pheromone or placebo for two months, after which the treatment groups were reversed. Five of the cat owners stated they saw some improvements in their pet's health, four didn't. Not very impressive. A systematic review of other studies not sponsored by the manufacturer reveals some decrease in urine spraying by some cats. In a veterinary hospital setting, there appeared to be no calming effect before surgeries or placement of catheters.

Our cat has long departed this world for cat heaven or, more likely given his destruction of the house, cat purgatory. He did live a long time, though, perhaps thanks to the constant exercise he got when I made him chase a laser beam up the walls. These new-fangled cat calming products certainly do arouse my chemical curiosity, but I'm not sure the stress of prospective cat pee can be countered by the satisfaction of carrying out an experiment on cat chemistry. I've actually had my share of cat pee experiments. Ridding the house of the smell of 3-mercapto-3-methylbutan-1-ol is a challenge. Some commercially available bacterial enzyme preparations are not bad, but I found a concoction made by mixing hydrogen peroxide, dishwashing detergent and baking soda worked better. That's pretty well the classic formula for eliminating skunk smell. And it is based on some interesting chemistry. The smelly substance belongs to the family of compounds known as thiols. Upon reaction with hydrogen peroxide, the thiols are converted to the far less fragrant sulfonic acids. These in turn can be neutralized by reaction with alkaline baking soda to form water-soluble salts that can be washed away by the detergent.

Interestingly, the facial cat pheromone mix is now available in a plug-in electric diffuser much like scented air fresheners.

The manufacturer claims that the spray can greatly reduce territorial spraying and stress-related behavior such as loss of appetite, anxiety or depression. I'm not sure how one diagnoses anxiety or depression in a cat, but I do know a little about cat-induced anxiety in a cat owner. Maybe it's the owner who needs a spray. And there seems to be one available. The promoters of 1Hour Break claim their oral spray concocted from extracts of lobelia, pulsatilla, St. John's Wort, passion flower and kava kava root reduces anxiety in humans. The evidence, however, is all anecdotal. A similar product, Rescue Remedy, contains flower extracts, including "impatiens to mollify impatience" and cherry plum, to "calm irrational thoughts." There seems to be one rather weak double-blind trial that "suggests Rescue Remedy may be effective in reducing situational anxiety." I might have given that one a try when we still had our cat, since he did cause me to have some irrational thoughts.

HEALTH MATTERS

COBWEBS IN THE NOSTRIL, MASHED MICE AND DANCING PATIENTS

"The desire to take medicine is perhaps the greatest feature which distinguishes man from animals." So claimed Sir William Osler, McGill University graduate, professor of medicine at McGill University, one of the founders of the Johns Hopkins University School of Medicine and often labeled as the "Father of Modern Medicine." Osler introduced the concept of clinical clerkship, insisting that third- and fourth-year medical students be exposed to hands-on experience with patients and also pioneered the idea of a residency program for medical graduates to further hone their skills. He reportedly asked for no other epitaph than that he taught medical students in the wards, something he considered his most important work. By putting less emphasis on lectures and books, and more on practical skills, he fundamentally changed the way medicine was taught. "Listen to your patient," he told his students. "He is telling you the diagnosis."

Osler's statement about man's desire to take medicine, which was made in a lecture he delivered to the public in 1891, is widely quoted, but his follow-up sentence isn't. "Why this appetite should have developed, how it could have grown to its present dimensions, what it will ultimately reach are interesting problems in psychology." He was expressing his concern that while physicians "have gradually emancipated ourselves from a routine administration of nauseous mixtures on every possible occasion, and when we are able to say that a little more exercise, a little less food, and a little less tobacco and alcohol may possibly meet the indications of a case . . . you, the people, should wander off after all manner of idols and delight more and more in patent medicines and be more than ever at the hands of advertising quacks."

It isn't surprising that Osler was critical of the use of medicines at the turn of the twentieth century because most of them did not do much good. "One of the first duties of the physician is to educate the masses not to take medicine," he maintained, believing in the self-limiting nature of disease. Osler himself prescribed relatively few drugs, his basic armamentarium consisting of quinine for malaria, digitalis for heart failure, opiates for pain and coughs, iron and arsenic for anemia. Instead of drugs, he recommended bleeding the patient for a variety of conditions including pneumonia, stroke and mumps, and suggested acupuncture for sciatica and neuralgia. If that didn't work, he favored trying aquapuncture, the injection of distilled water. He also thought that for nosebleeds there was no harm in trying the insertion of a cobweb into the nostril.

That of course sounds pretty curious to us, but throughout history people have resorted to every imaginable remedy for their ailments, from a toothache poultice made of mashed mouse to a whiff of flatulence stored in a jar to ward off the Black Plague. Most of the treatments failed, but eventually through trial and error, at the expense of much misery, some effective drugs did emerge. As early as 70 AD, Dioscorides, for example, described the use of the seeds of autumn crocus to treat gout. The active ingredient, colchicine, was not extracted and identified until 1820. Some drugs, penicillin being a classic example, were discovered by accident, others such as taxol for cancer by a meticulous search for physiologically active compounds found in nature.

Today, drug research focuses on molecular structure and known mechanisms of action. Gleevec (imatinib), a drug that dramatically increases the survival rate in chronic myelocytic leukemia (CML), was developed based on the finding that CML patients produce an abnormal version of the enzyme tyrosine kinase, which in turn leads to an overproduction of white blood cells. Knowing the molecular structure of the enzyme, researchers were then able to design and synthesize a compound that would inhibit its action.

In some cases, the effectiveness of a drug for an ailment was discovered when it was being used for a different condition. Antidepressants known as monoamine oxidase inhibitors (MAOI) are a classic example. In 1951, isoniazid was introduced as a treatment for tuberculosis with great success. But it wasn't long before concerns about bacterial resistance arose, and when that happens, chemists begin to tinker with the molecular structure of the drug to develop a derivative that will help stave off resistance. Within a year, iproniazid was ready for testing in tuberculosis hospitals. While it turned out not to be effective against TB, the drug had a remarkable side effect. Doctors and nurses noted a significant improvement in patients' mood, with some even taking to dancing in the hallways. Not a common sight in any hospital.

As it turned out, iproniazid inhibited the action of monoamine oxidase, an enzyme that normally degrades norepinephrine and serotonin, two compounds that control mood. Inhibition of the enzyme raises blood levels of both and leads to feelings of happiness. After favorable results were obtained on testing depressed patients, iproniazid hit the marketplace as Marsilid, only to be withdrawn in 1961 because of liver toxicity. But iproniazid had demonstrated the principle of antidepressant action and opened the way to the introduction of other monoamine oxidase inhibitors, which are still in use although they have mostly given way to the newer selective serotonin reuptake inhibitors (SSRIs).

Other drug actions that have been discovered in this fashion include sildenafil (Viagra), first introduced as an anti-anginal medication, however, it was its ability to elevate more than just mood that brought it fame and fortune. Minoxidil (Rogaine) hit the market as an antihypertensive but when patients began to show unusual hair growth, the drug's manufacturer seized the moment and converted it to a "hair regrowth treatment." The there's botulinum toxin, a substance that debuted as a treatment for crossed eyes and uncontrollable blinking before the accidental discovery of its antiwrinkle effect led to its use as the blockbuster Botox. Amphetamine was introduced as a treatment for congestion and asthma, but it was its stimulant side effect that led to its use both by the Allies and the Germans during World War II as a performance-enhancing substance. A further surprise was when amphetamine, despite being a stimulant, proved to be effective in the treatment of attention deficit hyperactivity disorder (ADHD).

So then, is it really the desire to take medicine that distinguishes us from animals, as Osler opined? I would suggest it is the making of medicines rather than taking them that sets us apart. After all, chimps have been known to seek out certain plants when they feel ill, but only humans have the ability to isolate and identify and perhaps improve upon the active ingredient. Who knows, maybe our next drug will come from some sort of monkey business.

PELLAGRA SLEUTH HELPED END THE "SCOURGE OF THE SOUTH"

He was known as the Sherlock Holmes of medicine. Dr. Joseph Goldberger forged a career as an investigator of disease outbreaks, but is best known for solving the puzzle of pellagra, ominously referred to as the disease of the four "d"s: dermatitis, diarrhea, dementia and death. This "scourge of the South" erupted into an epidemic in 1906 and struck some 3 million people by the time it began to decline in 1940. By that time it had dispatched about 100,000, mostly poor southerners. The swath of death would have continued had it not been for the epidemiological skills of Joseph Goldberger.

Born in 1874 in Hungary, Goldberger shared both the country and year of birth with another tenacious investigator, Harry Houdini. While Houdini probed the possibility of life after death, Goldberger focused on keeping the living in this world as long as possible. Joseph was brought to the U.S. at the age of seven by his parents, who were in search of the American dream. The tenements of New York turned out to be more of a nightmare; nevertheless his grocer father managed to send his son to college. After two years of engineering, young Joseph switched paths and in 1895 graduated as a physician at the top of his class.

General practice turned out to be somewhat unfulfilling and Goldberger, craving travel and adventure, took an exam for a commission in the U.S. Public Health Service and ranked first among the applicants. His career as an epidemic fighter was under way. It was tough going, and Goldberger contracted almost every disease he investigated: dengue fever in Texas, typhus in Mexico and yellow fever in New Orleans. But in New

Orleans he caught something else as well, the eye of debutante Mary Farrar. That was a real problem with Joseph's orthodox Jewish parents objecting to their marriage as strongly as Mary's Episcopalian family.

In 1914, the Surgeon General asked Goldberger to undertake an investigation of pellagra, a problem to which he would devote the rest of his life. At the time pellagra was thought to be an infectious disease, and patients were often quarantined and their families ostracized. Since the disease was common among cotton pickers, some targeted cottonseed oil as the cause. Nonsensical cures ranging from arsenic and castor oil to strychnine and spring water baths were proposed.

It wasn't long before Goldberger published his first paper, "Etiology of Pellagra; The Significance of Certain Epidemiological Observations with Respect Thereto," in which he noted that in mental hospitals and orphanages where pellagra among inmates was rampant, staff members did not contract the disease. Goldberger had worked with infectious diseases enough to know that germs did not distinguish inmates from employees and suggested that "the explanation of the peculiar exemption under discussion will be found in the opinion of the writer in a difference in the diet of the two groups." And so it would turn out to be.

But proposing a hypothesis and proving it are two different matters. Goldberger set out to show that pellagra could be cured by altering the diet and that it could be induced with a restricted diet. His first experiment took him to an orphanage where half the orphans had pellagra. In one ward he instituted a diet that included milk, meat, eggs and fresh vegetables, while in another ward the orphans were kept on the usual diet of the South, basically grits, corn bread, collard greens, fatback and molasses. Pellagra was wiped out in the treated group, but its incidence was unchanged in the control ward.

Goldberger's proposal to prove that pellagra could be induced was more controversial, since trying to make people sick does not exactly mesh with the ethics of medicine. However, at the time prisoners were sometimes offered freedom if they "volunteered" for a medical trial. Eleven murderers, embezzlers and forgers were enlisted with aid of the Governor of Mississippi and agreed to eat only the corn-based diet Goldberger had approved. The results were clear. After just six months, five of the men had contracted pellagra. Goldberger was accused of torture, and critics complained that the whole experiment had been set up to pardon two convicts who were friends with the governor.

Goldberger responded by organizing "filth parties" in which skin, nasal secretions, urine and feces from pellagra patients were mixed into dough and swallowed by volunteers. There was some nausea and diarrhea, but no pellagra. It simply was not an infectious disease.

The challenge now was to find the exact nature of the "pellagra preventive factor." Goldberger went onto show that brewer's yeast prevented the disease as effectively as milk, meat and vegetables, but never managed to identify what the responsible component was. In 1937, Conrad Elvehjem extracted niacin from liver and soon after Tom Spies demonstrated that the compound cured pellagra. Niacin fell into the category of vitamins, since these are substances that must be included in the diet to prevent specific diseases. Vitamin B3, as niacin came to be known, can actually be made by the liver from tryptophan, a common amino acid in proteins. But conversion is slow and it takes about 60 mg of tryptophan to make 1 mg of niacin. So unless someone has an extremely high protein diet, we need to supplement other foods with niacin. Goldberger's enthusiasm for recommending changes to the diet of the poor, and the addition of niacin to bread starting in 1938, led to a dramatic decline in pellagra.

Curiously, the disease had not been noted in Mexico in spite of much of the population living on corn as a staple. That's because niacin is actually found in corn but is bound to other molecules in a way that makes in unavailable to the body. In Mexico, corn is traditionally soaked in a lime solution before cooking, a process that releases the bound niacin.

Today, at least in the developed world, pellagra is no longer an issue, but niacin still makes it into the news. It turns out that in doses far greater than that needed to prevent pellagra, it has a beneficial effect on blood cholesterol profile, although it is associated with a flushing of the face that many find intolerable. Joseph Goldberger's dogged work in tracking down the cause of pellagra is a masterpiece of investigative craftsmanship, certainly worthy of Sherlock Holmes.

A HEALTHY TAN? FORGET IT!

When it comes to health matters, scientists rarely make statements that do not begin with "may." But here is one: excessive exposure to sunlight causes skin cancer. There's no "may" about it. And here is another one. Chemical protection can effectively reduce exposure. Uncertainties do, however, emerge when it comes to deciding on which specific chemicals to use. Activists claim that some sunscreens are unsafe and blame regulatory agencies for not looking after the welfare of the public while manufacturers profess that their products have been thoroughly tested for safety and efficacy. Actually, when you blow away the superfluous blather emanating both from the alarmists and from industry, there is some simple advice to offer.

The challenge is clear. Find a chemical or mixture of chemicals that can be applied to the skin to reduce exposure to the full spectrum of ultraviolet light. Then make sure these chemicals do not degrade upon exposure to light, have no topical or systemic toxicity, are minimally absorbed into the body, are resistant to water, do not have a greasy feel, are cosmetically acceptable, do not stain clothing and can be incorporated into a vehicle that allows for easy spreading. Quite a list of demands.

The first commercial "sunscreens" appeared in the 1960s and were designed to filter out UVB, the shorter wavelengths of ultraviolet light (290 - 320 nanometers). These are the rays that cause sunburn, which was the main concern at the time. Slightly longer waves, those responsible for tanning, were deemed safe. Finding chemicals that absorb the nasty UVB rays was not particularly difficult, with para-aminobenzoic acid (PABA), octocrylene, phenylbenzimidazole sulfonic acid and various cinnamates and salicylates being up to the task.

Products with different concentrations of these ingredients were introduced for different skin types, each prominently featuring a Sun Protection Factor (SPF), basically a measure of the time it takes for skin to redden compared with having no protection. The SPF value is determined in the laboratory by applying 2 mg of product per square centimeter to the skin of volunteers. Using a product with an SPF of 15 means that a person who normally begins to burn in ten minutes can in theory stay in the sun for 150 minutes before experiencing any visible effect on the skin.

It didn't take long for this scenario to prove to be too simplistic. As a clear link between skin cancer and UVB emerged, the focus shifted from preventing sunburn to preventing skin cancer, resulting in an industry frenzy of products with higher and higher SPF values. In truth, an SPF of 15 already blocks 94 percent of UVB, only 3 percent less than one labeled as SPF 30. In any case, these numbers are only meaningful if the product is applied the same way as in the lab studies, which turns out not to be the case. Most people were applying far less than 2 mg per square centimeter and were not getting the protection they thought they were getting. What many were getting, though, were various skin reactions. And something else became apparent as well. The longer wavelengths of ultraviolet light, 320 - 400 nm, known as UVA, previously thought to be innocuous, were found to be more deeply penetrating than UVB and responsible for premature wrinkling and aging of the skin (photoaging). Unlike UVB, they can even pass through glass. Furthermore, UVA also was found to be potentially carcinogenic.

Now there was a need for a novel class of products that would protect the skin both from UVB and UVA. Ideally, not one that would just absorb some wavelengths, but one that would reflect all ultraviolet light. Titanium dioxide and zinc oxide, both mineral pigments, fit the bill, but left a white residue on the skin. That was all right for lifeguards' noses, but not for vain sunbathers. The search was on for cosmetically acceptable molecules capable of absorbing UVA. Oxybenzone and avobenzone (Parsol 1789) were up to this task, but as usual, there are some "buts."

When oxybenzone absorbs ultraviolet light it becomes energized, and some of this energy is dissipated through the production of free radicals. These are very active molecular species that have been linked to cancer. Oxybenzone also undergoes a reaction in the presence of ultraviolet light to form a compound called a semiquinone, which in turn can inactivate some of the naturally occurring antioxidants in the skin, such as reduced glutathione. Not a good thing, since antioxidants offer protection against free radicals. And if that weren't enough, it turns out that oxybenzone can also mimic the behavior of estrogens, at least in fish exposed to high doses. It has therefore been labeled a potential endocrine disruptor. Concern has been raised, mostly by the Environmental Working Group, an American activist organization, because surveys have shown that oxybenzone can be found in the blood of 97 percent of the population.

But, and a big but it is, there is no evidence reported in the scientific literature of oxybenzone being linked to any human health problem, except for photodermatitis, a skin reaction triggered by exposure to sunlight. There are hundreds and hundreds of compounds, both natural and synthetic, that if scrutinized the same way as oxybenzone, could be linked to problems. Phthalates, bisphenol A, soy extracts and various pesticides are estrogenic. We live in a world full of hormone-like substances and a complete analysis of our blood would reveal hundreds of these. All of this goes to say that the risks of oxybenzone as implied by the Environmental Working Group, I think, are overstated.

Avobenzone is cosmetically elegant and non-irritating, but becomes unstable after a couple of hours of exposure to ultraviolet light. However, its stability is increased when combined with oxybenzone, especially if another stabilizing agent known as diethylhexyl–2,6–naphthalene (DEHN) is added. This combination, developed by Neutrogena, is known as Helioplex. An important question arises here. What happens to the UV energy that these chemicals absorb? The energy has to go somewhere — might it not have a damaging effect? DEHN takes the energy absorbed by avobenzone and transfers it to oxybenzone, which then fluoresces it as harmless red light.

Another effective broad-spectrum sunscreen is terephthalylidene dicamphor sulfonic acid, which goes by the trade name Mexoryl. It is stable, absorbs UV light and dissipates the energy as harmless heat. Mexoryl isn't absorbed through the skin and so far there are no safety issues. And recently, excellent products using micronized titanium dioxide and zinc oxide have been developed that do not leave a telltale white residue. Presently it is difficult to judge exactly how much protection a product affords against UVA because no SPF-like system has yet been devised. But regulatory agencies are working on it.

There is one more "may" about sunscreens that has been converted to fact. We no longer have to say that sunscreens may prevent skin cancer, we can say they do. A study in Australia, where skin cancer is a huge concern, involved 1,600 subjects who were given sunscreen to use every day for four and a half years. They developed 40 percent fewer squamous cell cancers than a control group who just maintained normal skin care without being given specific instructions about the use of sunscreens.

So there it is. Sunscreens can prevent skin cancer, which is not a rare disease. The World Health Organization estimates 48,000 deaths a year from melanoma (likely sun related but not conclusively proven) and 12,000 from other forms of skin cancer. What to do?

Look for a product with SPF 30 containing avobenzone, Mexoryl, titanium dioxide or zinc oxide. Apply fifteen minutes before going out in the sun; use a shot glass full for the body and half a teaspoon for the face. Reapply frequently. Forget terms like "waterproof," "all day protection" and "sweatproof." They're meaningless. And if you are buying something that is "chemical-free," you are not getting a good deal because you're buying a vacuum. Sunscreens should not be used to prolong sun exposure but rather to protect the skin when exposure is unavoidable. Above all, remember that unfortunately there is no such thing as a healthy tan.

CAT BLOOD, DOG LIVER AND A DOSE OF LUCK

Jay McLean walked into Dr. William Howell's laboratory at Johns Hopkins University and placed a beaker of cat blood on his desk. "Can you please tell me when the blood clots?" he asked. It never did. And that striking observation in 1916 turned out to be momentous in the history of science, opening the door to the discovery of heparin, the anticoagulant drug that dramatically altered the practice of medicine.

McLean was a medical student who had not exactly followed the beaten track. He had a tough childhood, losing his father when he was four years old and his home to the San Francisco Fire when he was fifteen. During his last year of high school, young McLean got his hands on a book about medical education and was smitten. But entry into medical school required an undergraduate degree, which McLean decided to pursue at the University of California at Berkeley. It was here that he discovered a textbook on physiology written by William Howell, a professor of medicine at Johns Hopkins University in Baltimore. This was captivating stuff, and McLean decided that Johns Hopkins Medical School was the place for him. But there was a problem. His application to Hopkins was denied.

Nevertheless, McLean decided to uproot himself and take a shot at talking his way in, and if that didn't pan out, he'd work for a year and then try again. After managing to wangle an appointment with the dean, he was told that there was no recourse once an application had been denied. But then fate intervened. Word came the next day from the dean about an unexpected vacancy. McLean would be admitted to medical school as long as he made up a missing requirement, a lab course in organic chemistry.

An overjoyed McLean then approached Dr. Howell and

described how the professor's text had triggered an interest in research in physiology, which he believed would prepare him for a career in surgery. He now had a year to show that he could accomplish something significant by himself! The professor apparently liked the young man's bravado and assigned him a project dealing with the coagulation of blood, his own area of interest. Howell had previously shown that a crude extract of brain tissue promoted blood clotting and suspected that the active ingredient was cephalin, a complex fatty substance belonging to a family of compounds known as phosphatides. McLean was to try to prepare cephalin in a pure form.

Working with macerated brain tissue that produced an insufferable odor turned out to be an unpleasant task. Worse, cephalin seemed to defy all attempts at isolation in a pure form. But McLean learned that German researchers had found heart and liver extracts with coagulant activity and figured that these might lead to an easier purification of cephalin. They didn't. And then chance intervened once more. McLean decided to retest one of his phosphatide samples that had been stored for a while and discovered that not only had the coagulant activity been lost, it had been replaced by an anticoagulant effect!

Howell's skepticism of the discovery prompted McLean to resort to the cat blood demonstration. But as his year of research was coming to an end, and medical school was calling, McLean regrettably had to abandon any further experiments. However, the idea of isolating an anticoagulant had been planted in Dr. Howell's mind, and within a couple of years he managed to isolate a water-soluble substance from the liver of a dog that in the laboratory had reproducible anticoagulant activity. Heparin, as he would name the extract from the Greek word for liver, had basically a carbohydrate structure, chemically very different from the fat-soluble cephalin.

Surgeons had long been plagued by the formation of blood clots as their scalpels damaged tissues. Also well known was the role of blood clots in heart attacks, pulmonary embolism and deep vein thrombosis. Any medication capable of preventing blood from clotting, in other words, an anticoagulant, would be most welcome! But the crude heparin preparations resulted in patients developing a variety of side effects, ranging from fever and chills to shock. Solving this problem appealed to Dr. Charles Best, who in 1929 had taken up a post as professor of physiology at the University of Toronto.

Best, already famous for having assisted Frederick Banting in the discovery of insulin, headed a group that eventually managed to purify heparin. Important contributions were also made by a team of Swedish scientists under Erik Jorpes, who had learned about heparin when he visited Best. Human trials began in 1935, and by the end of the decade, heparin had entered clinical practice and remains an indispensable medical weapon.

It took a while for the molecular structure of heparin to be determined, with many scientists playing important roles including McGill Professor Arthur Perlin, who just happened to be my Ph.D. research director. Heparin, as extracted from pork intestines or beef lung, is now known to be a family of carbohydrate polymers varying in molecular weight. Chemical processing methods have been developed to break down the long carbohydrate chains to yield "low molecular weight heparin" (Lovenox), which is widely used to treat deep vein thrombosis, prevent pulmonary embolism and reduce the risk of blood clot formation associated with a heart attack. No emergency room is without it.

Although the term "wonder drug" is often recklessly bandied about, it isn't too far off the mark when it refers to the substance without which cardiac surgery or dialysis would be impossible.

Over the years, there has been controversy about who should get credit for the discovery. Should it be McLean, who first noted the anticoagulant effect, or Howell who managed to isolate the first sample of heparin? Most accounts lean toward Howell. Dr. Hans Selye, known around the world as the father of stress research, suggested that in science credit should be attributed to the individual whose work has led directly to the development and application of the discovery. Since Jay McLean's discovery led directly to Howell's development of heparin, McLean should therefore get the credit. Although perhaps one could also make an argument for the gallant cat that donated the blood that stimulated the research leading to heparin.

FROM THE MANCHINEEL TO THE YEW, TREES ARE A SOURCE FOR GOOD OR EVIL

According to Talmudic tradition, a wise rabbi once proclaimed that if a person planting a tree were told that the Messiah had arrived, he should finish planting before going to greet him. That of course was long before there was any awareness of the important role trees play in generating oxygen and soaking up carbon dioxide. Neither was there any knowledge about trees preventing soil erosion, absorbing air pollutants, preventing water runoff or controlling climate by moderating the effects of the sun, rain and wind. Nevertheless, the importance of trees was clear. They produced olives, pomegranates, nuts, dates and figs while providing wood for cooking and timber for building.

As time went on, trees would prove to be a source of important commodities such as paper, rubber, maple syrup, cork, turpentine, amber, tannins, fragrances (eucalyptus, sandalwood), flavorings (cola, cinnamon, bay leaves, nutmeg) and medicines (camphor, quinine, paclitaxel, salicin). Little wonder that environmentalists who appreciate the value of trees are referred to as tree-huggers. But they do need to be careful about which trees they hug, with the manchineel tree being a particularly poor choice. In this case, expressing affection can prove to be a blistering experience. Manchineel trees, native to the Caribbean, Central America and Florida, produce a variety of toxins, phorbol being an example, that are not only irritating to the skin but can produce severe illness when ingested. Biting into the tree's fruit, which resembles an apple, can cause dramatic swelling of the airway and severe gastroenteritis.

Legend has it that some Caribbean tribes would tie prisoners to the trunk of the manchineel tree as a form of torture, subjecting the unfortunate captives to terrible blistering as rain washed the tree's milky sap over their bodies. It seems that concentrated manchineel sap can even kill! Juan Ponce de León, the Spanish explorer who led the first European expedition to Florida, can attest to the lethal effects of the manchineel's poison. The explorer and his men were attacked by natives as they sought to colonize the land he named after its floral landscape, searching for gold, not for the fountain of youth, as is commonly believed. During the scuffle, Ponce de León was struck in the thigh with an arrow that was apparently poisoned with the sap of the manchineel tree. He managed a retreat to Cuba, where he died of his wounds.

Some African tribesmen tip their arrows or spears with an extract of the bark of the *Acokanthera schimperi* tree. Ouabain, the toxic ingredient is a sodium/potassium ion pump inhibitor capable of stopping the heart in its tracks. Even that of an elephant! But the African crested rat is immune to the toxin and actually uses it to its advantage. The rat chews on the bark of the tree and then licks the fur on its back. Any predator that looks on this foot-long rat as a tasty morsel ends up making a fatal error when it grabs its prospective meal. Why the rat is unaffected by ouabain is a mystery, the solution of which may have implications for the treatment of heart disease in humans.

Perhaps the best-known tree poison is strychnine, found in the seeds of the species *Strychnos nux-vomica*, already familiar to the ancient Egyptians. Legend has it that Cleopatra used her servants as guinea pigs, forcing them to swallow the poisonous seeds when she was contemplating her own suicide. Upon witnessing their agonizing death, she decided that the bite of an "asp," likely an Egyptian cobra, was a better way to go. Some historians, however, suggest that Cleopatra actually made her exit with a mixture of hemlock, wolfsbane and opium.

While Cleopatra did not end up using a tree poison, Boudica, the warrior queen of the ancient Britannic Iceni tribe, did choose to ingest the poisonous leaves of the yew tree, preferring death to being taken prisoner by the Romans after losing a battle. The English yew, or *Taxus baccata*, contains taxane, an alkaloid that can cause cardiovascular collapse by interfering with calcium and sodium channels in heart muscle cells. There's no way to confirm the story of Boudica's suicide, but a case report in a 2011 issue of *Clinical Medicine* does describe a seventeen-year-old girl's attempted suicide with yew leaves. Apparently she changed her mind after swallowing the toxic leaves and sought help in a hospital's emergency room. After being admitted, her heart began to beat very erratically, necessitating defibrillation and the administration of the anti-arrhythmic drug amiodarone. After recovery, she admitted being taken by the story of Boudica, which she had encountered in her ancient British history class.

A young man who was trying to carry out a mystical ritual to experience the afterlife through the use of yew leaves was not so lucky. He had read about some ancient practice of consuming yew seeds when Venus and Jupiter came into alignment, allowing the indulger to experience the afterlife and return. We'll never know what he experienced because he did not return. The coroner's verdict was cardiac arrest caused by ingesting yew seeds.

While the yew tree's toxins can take a life, they can also save one. Back in the 1960s, after screening thousands of plants, a compound derived from the bark of the Pacific yew was found to be effective in the treatment of cancer. However, there was a problem. The isolation of paclitaxel required large amounts of bark, and stripping the bark killed the tree. Chemists quickly went to work to find a synthetic method to produce the compound but were stymied until they discovered a precursor in the leaves of the English yew that could be converted into paclitaxel. Since the collection of fresh clippings did not damage the tree, a supply of paclitaxel for the treatment of cancers of the lung, ovaries, breast, head and neck became available.

So, are the chemicals found in the leaves of the yew species good or evil? Neither. It's a question of how they are used. Chemicals don't make moral decisions, but people do. Maybe a bite of the forbidden fruit picked from the tree of knowledge in the Garden of Eden wasn't such a bad idea.

THE CHALLENGE OF MALARIA

In 1942, American and Japanese troops were locked in two bitter battles on the South Pacific island of Guadalcanal. Not only were they locked in a struggle for control of the island, they also had to wage war against the scourge that was malaria. The ancient Romans believed the disease was caused by vapors emanating from swamps and had called it "malaria" or "bad air." They weren't completely wrong. Wetlands do play a role, but not by releasing any sort of noxious vapors. Swamps are ideal breeding grounds for the mosquito that preys on humans and transmits the malaria-causing parasite.

No treatment for malaria was available until Jesuit missionaries learned about the fever-reducing properties of the bark of the cinchona tree from South American natives. Quinine, the active ingredient in cinchona bark, was isolated in 1820 by two French pharmacists, but supply could not meet demand. Attempts to synthesize quinine in the laboratory failed — no surprise given the molecule's complex structure. But a clue toward developing synthetic antimalarials emerged in the 1880s when German physician Paul Ehrlich noted that a dye called methylene blue was particularly effective in staining the malaria parasites.

Microbes, recently identified as disease causing agents, were notoriously difficult to see under the microscope. Ehrlich, however, discovered that they could be readily visualized when stained with some of the novel synthetic dyes that were being produced by the chemical industry. Interestingly, these dyes had a connection to quinine. Back in 1856, young William Henry Perkin had attempted to synthesize quinine from coal tar derivatives with no success, but one of his failures resulted in the synthesis of mauve, the world's first synthetic dye. This launched the dye industry, with methylene blue being one of its first successes. Ehrlich reasoned that since the dye reacted so readily with the malaria parasite, it might poison it if administered to a malaria victim. In 1891, he actually managed to cure two malaria patients with methylene blue, becoming the first doctor to ever administer a synthetic drug to a human.

Methylene blue was not always effective, and the search was on for an improved version. Using the compound as a prototype, chemists at the Bayer Company in Germany developed plasmoquine and mepacrine, both of which were effective. Then in 1934 they synthesized resochin which worked well but was thought to be too toxic. Both plasmoquine and mepacrine (Atabrine) were widely used by both Germans and Americans during World War II.

During the war, American, British and Australian scientists tested more than 16,000 synthetic compounds to find more effective, and safer antimalarials. Resochin, which had been renamed "chloroquine," was one of the compounds tested. At first, like the earlier German researchers, the Allied scientists rejected it as too toxic, but because of numerous other failures they decided to test it more extensively. By the time the results were in the war was over, but chloroquine would prove to be an extremely effective antimalarial, destined to become a mainstay of malaria treatment. Eventually, as was to be expected, the parasite developed a resistance to the drug, and synthetic chemists had to come up with newer analogues, which they did. But the synthetics were eventually joined by an effective naturally occurring compound.

During the Vietnam War, the North Vietnamese army was ravaged by malaria. China supported the North, and Chairman Mao Zedong ordered a research effort to find a treatment based on Traditional Chinese Medicine (TCM). A treatment based on

TCM would reduce illness among the Vietnamese and at the same time show the world that ancient Chinese methods outperformed drugs developed by the capitalists. The problem, though, was that the number of substances listed in traditional Chinese documents was immense, with hundreds of malaria treatments being described. But one preparation, described in 1596 by Li Shizhen, a famous herbalist, seemed particularly interesting because of its apparent success in reducing fever. It was an extract of the sweet wormwood (*Artemisia annua*), a plant that had been used for more than 2,000 years by Chinese herbal medicine practitioners, initially for the treatment of hemorrhoids.

In 1972, researcher Tu Youyou showed that an ether extract of the leaves of sweet wormwood was extremely effective at destroying the malaria-causing parasite in the blood, and by the late 1970s the active ingredient had been isolated and its molecular structure determined. There were no ethics committees to bother with in China during the heyday of Mao's rule, and artemisinin, as the compound had been christened, was tested on Chinese soldiers who were purposely infected with the parasite. It was effective, but there were reports of neurotoxicity, probably because large doses had to be used due to the compound's lack of solubility. A large research effort was then mounted to develop more soluble artemisinin derivatives, and by the 1990s artemether and artesunate, both synthesized from artemisinin were developed as effective oral, rectal and intramuscular antimalarials. These drugs treat but cannot prevent infection.

Most of the wormwood plants used to produce artemisinin needed for the production of the various derivatives are grown in China and Vietnam, but genetic engineering techniques may eventually supplant the need to rely on plants. Genes isolated from the wormwood plant that code for various enzymes used in the biosynthesis of artemisinin can be inserted into yeast cells, which then crank out the compound. Unfortunately there is also a flood of fake artemisinin medications that either contain no active ingredient or an ineffective amount. The latter is a huge problem, because while not offering any benefit these fakes can still trigger parasite resistance.

Back in 1942, neither artemisinin nor chloroquine were yet available, and Atabrine had a variety of side effects. There was a feverish effort in the U.S. to come up with a better drug, and at one point about 400 prison inmates in Chicago were infected with malaria in order to test potential cures. Whether or not the prisoners were given details about what was being done is questionable, but effective drugs such as primaquine did emerge from the research. During the Nuremberg Trials, Nazi doctors referred to this episode to try to justify their experiments on prisoners.

Unfortunately malaria still continues to be a scourge, affecting an estimated 200 million people every year and killing about half a million. More research is desperately needed, but hopefully it won't take another war to stimulate it.

PARSLEY, SAGE, ROSEMARY AND THYME

In the Middle Ages, the town of Scarborough in Yorkshire, England, featured an annual fair that attracted merchants from all over the country as well as from the Continent. An array of fabrics, dyes, skins, pots and foods vied for customers' attention. And then there were the herbs. There would have been a large assortment, but surely parsley, sage, rosemary and thyme would have been among them. After all, Simon & Garfunkel told us so in the lyrics of "Scarborough Fair," the memorable ballad featured on the soundtrack of the classic movie *The Graduate*.

While Simon and Garfunkel catapulted the song to fame, various versions of the melody and lyrics can be traced back to the seventeenth century. Some historians claim that these specific herbs were mentioned both because of their medicinal properties and the mystical belief at the time that herbs had the ability to influence emotions. Parsley, for example, was thought to remove bitter feelings the same way it eliminated bad odors. Chewing fresh parsley was a long-standing antidote to bad breath. The botanical name of sage, *Salvia officinalis*, derives from the Latin "*salvere*," meaning "to be saved" and pays homage the Roman belief that the herb was a key to longevity. In the Middle Ages, sage was actually one of the components of a concoction known as "Four Thieves Vinegar" that claimed to offer protection against the plague. It didn't.

Rosemary was also part of that potion, but historically the herb is better known for its supposed memory-enhancing effect. In ancient Greece, so the story goes, students would hang rosemary around their neck to improve memory and concentration. That actually may have worked had they also prepared for their exams while sniffing rosemary. Modern studies have shown that recall is improved when subjects are exposed to the same smell during a test as during the learning process. The strong, lingering scent of rosemary may well have been responsible for its inclusion in medieval wedding bouquets as a symbol reminding lovers of their vows. Thyme also has a long-lasting and pleasing scent that was thought to ward off melancholy. The ancient Greeks placed some in their baths.

There was also a more practical reason for sale of these herbs. Microbial contamination of food was a scourge at the time, and many herbs and spices are known to contain compounds with antimicrobial activity. Thyme oil, for example, is being explored today for its antibacterial effect, particularly against *Listeria monocytogenes*. On top of being effective against bacteria, thyme oil can be labeled as a natural preservative, a strong selling point. Thymol, the major active ingredient, also has potent antioxidant properties and can prevent fat from becoming rancid. Rosemary extract also contains the antioxidants carnosic acid and carnosol and has been approved for use in meats, baked goods, oils and fish-oil supplements. Curry may well have developed as a popular flavoring because of the antibacterial effects of turmeric, coriander and nutmeg.

Vendors at Scarborough Fair would surely have been hawking more than just parsley, sage, rosemary and thyme. There would have been mugwort to ease labor pains, burdock and savory to help pass flatulence, cottonweed for headaches and, in the words of Nicholas Culpeper, the prime authority on herbalism at the time, foxglove to "purge the body both upwards and downwards, sometimes of tough phlegm and clammy humours, and to open obstructions of the liver and the spleen." Culpeper was a botanist, herbalist physician and astrologer who forged a system of treatments that mixed reasonable use of herbs with nonsensical "medical astrology."

There was also belief in the doctrine of signatures, which maintains that nature provided humans with clues about the treatment of disease, so plants or herbs that resemble parts of the human body were used to treat ailments of that part of the body. Lungwort, for example, would help with disorders of the lung, bloodroot for diseases of the blood and beans were of help with kidney problems. Indeed, the history of herbal medicine is characterized by a curious blend of science and nonsense. Not too different from today. Just consider oil of oregano, with its claims to treat sore throats, lice, colds, acne, infections, parasites, yeasts, diabetes, allergies or whatever one fancies.

No less an authority than Dr. Oz devoted a segment of his show to explaining how carvacrol, the "super ingredient" in oil of oregano, destroys nasty bacteria and boosts the immune system. There was even a neat demo, in which a vile-looking model of a bacterium was encased in what looked like a glass bubble. Dr. Oz attacked the bubble, which played the role of the bacteria's protective layer, with a kitchen knife. The attack wasn't exactly a challenge to the famed *Psycho* shower scene, and was not successful. Then Mrs. Oz stepped in with a kettle of hot water, which played the role of carvacrol, and poured it over the bubble. It immediately cracked and her knife-wielding hubby now easily burst through and punctured the bacterium, deflating it like a balloon. A really neat demo. I think they must have cooled the glass first to make it crack so easily. They get points for that one. Of course the point is overhyped. There is some cursory laboratory evidence of oil of oregano having an antibacterial effect: when bacteria are bathed in the oil, they perish. Of course they also perish if bathed in a salt solution, alcohol, lemon juice or a variety of soft drinks. It isn't hard to kill bacteria in a petri dish. But the body is not a large petri dish.

There is no evidence that a dose of oil of oregano is absorbed into the bloodstream to an extent where it may have an antibacterial effect. What about its claimed immune-boosting property? Here the evidence comes from nursing pigs. If they are given oil of oregano, they produce somewhat more white blood cells in their milk. Hardly something to oink about. What we have here are a few studies that suggest an effect in the lab or in animals that are then over-interpreted by marketers. Perhaps just like the overinterpretation of parsley, sage, rosemary and thyme. Maybe those particular words just had the right cadence and rhyme to fit the song.

O SLEEP! O GENTLE SLEEP

"O sleep! O gentle sleep!
Nature's soft nurse, how have I frightened thee,
That thou no more wilt weigh my eyelids down,
And steep my senses in forgetfulness?"

Henry IV, Part II is not one of the Bard's most memorable plays. I think it once lulled me to sleep. But these lines speak of insomnia, a common problem that begs for a solution. There is no shortage of advice. Count sheep. Drink warm milk. Feast on turkey. Take melatonin pills. Take kava-kava. Try valerian root. Mix up a drink from a special powdered blend of pumpkin seeds and dextrose. Listen to recordings of chirping crickets. Settle down on a mat embedded with amethyst crystals. Relax on a Polar Power Mega-Field Slumber Pad designed by Dr. William Philpott, whose last name rhymes with a term that can be used to describe his ideas about treating disease.

Virtually all diseases, Philpott maintained before he left us, could be managed or reversed with magnet therapy. Of course you had to have the right type of magnet. Only those that were capable of producing a "negative magnetic field" were therapeutic since "only these can promote an oxygen-alkaline rich environment within the body." That environment doesn't come cheap. Philpott's miraculous pads are still being sold for hundreds of dollars. But instead of focusing on the claptrap of negative magnetic fields, let's look at something that may actually have a positive effect. Like that mixture of pumpkin seed powder and dextrose.

First we need to do a little traveling back in time to the 1970s and the lab of MIT neuroscience professor Richard Wurtman. Unlike Philpott's random ramblings, Dr. Wurtman's research is backed by hundreds of peer-reviewed publications that have established him as one of the world's leading authorities on chemical activity in the central nervous system. It was Wurtman who demonstrated that levels of the neurotransmitter serotonin in the brain respond to dietary manipulation. This is important because higher serotonin levels have been linked with anti-anxiety effects, appetite suppression and sleep enhancement.

Serotonin is formed inside cells from the amino acid tryptophan, a component of most dietary proteins. When some questionable info emerged about turkey containing high doses of tryptophan, the lay press was ready to jump. Turkey became a remedy for insomnia and even made an appearance on a *Seinfeld* episode in which Jerry and George conspire to put Jerry's current girlfriend to sleep by overdosing her on turkey so that they can play with her collection of antique toys.

Actually, turkey protein does not have more tryptophan than other meat proteins. In any case, as Wurtman demonstrated, tryptophan levels cannot be increased by eating more protein. That's because amino acids are ferried across the blood-brain barrier by transporter molecules that have less of a preference for tryptophan than for the other amino acids that make up proteins. However, should a tryptophan-containing food be coupled with a source of carbohydrates, levels of tryptophan in the brain, and consequently serotonin, will rise. This happens because carbohydrates stimulate the release of insulin, which prompts the absorption of amino acids into muscles. But here too tryptophan is absorbed less efficiently, meaning that with the competing amino acids being driven into muscles, more tryptophan is available for absorption into the brain. Eating a turkey sandwich, with the bread providing the required carbs, actually makes some sense.

While serotonin may have a calming effect, it doesn't actually induce sleep. But the hormone melatonin does! And it is made in the brain's pineal gland from serotonin. This reaction, however, is inefficient as long as the eyes are stimulated by light. But with darkness, conversion of serotonin to melatonin begins and drowsiness sets in. The formula for sleep would then appear to be coupling darkness with a source of tryptophan and a carbohydrate that stimulates quick insulin release.

Wurtman's research prompted Canadian psychiatrist Dr. Craig Hudson to investigate the possibility of a commercial product designed to increase melatonin levels. He knew that melatonin supplements were available, but evidence indicated that when taken in a pill form, the hormone has a short half-life. Hudson's idea was to try to induce a normal sleeping pattern with a more continuous release of melatonin. First, he needed a good source of tryptophan and found it in pumpkin seeds. He then mixed the powdered seeds with glucose, the archetypical insulin releaser. A bit of natural lemon or chocolate flavor, and Zenbev sleep-enhancer was born. It hit the market after a double-blind placebo-controlled clinical trial showed that subjects with sleep problems were able to reduce the time spent awake during the night. Admittedly, a single study is not very compelling, but there seems to be no risk giving Zenbev a shot.

Neither is there a risk, outside of a possible allergy, in eating two kiwifruits an hour before bedtime. That's right, kiwis may help with sleep problems! In a study of twenty-four subjects, sleep onset, sleep duration and sleep quality were significantly increased with kiwi consumption. But why study kiwis at all in this context? It turns out that the fruit is a source of serotonin. Although the authors declare no conflict of interest, they do acknowledge support from Zespri International Limited. A quick Google search reveals that Zespri is a marketer of kiwifruit. That of course does not invalidate the study, but it would be comforting to see the trial duplicated by a totally objective research group. In the meantime there's no harm in giving the kiwi regimen a shot. Serotonin aside, kiwis are a great source of antioxidants and folate.

And if Zenbev or kiwis don't lull you to sleep, you can indulge in a cup of decaffeinated Counting Sheep Coffee. It contains valerian root extract, which does have a history of use as a sedative. During World War II, it was even used in England to relieve the stress of air raids. But as far as this coffee goes, we just have to take the marketer's word for its sleep-inducing effect. That, though, coupled with an appearance on television's *Dragon's Den*, seems to have been enough to perk up sales. And that should make the investors in Counting Sheep Coffee sleep better.

A SCOURGE TO MORE THAN CRUISE SHIPS

It's the scourge of cruise ships. It used to be referred to as Norovirus, but now the International Committee on Taxonomy of Viruses strongly encourages that the term be replaced by the term Norwalk virus. "*Noro*" apparently is a common name in Japan and there is concern that children with this name would be teased for "being a virus." But there are of course more important reasons to be concerned about the virus that was named after Norwalk, Ohio, the town where an outbreak of acute viral gastroenteritis sickened children at the Bronson Elementary School in November 1968.

The term "gastroenteritis" derives from the Greek "*gastro*" for stomach, "*entero*" for small intestine and "*itis*" for inflammation. So gastroenteritis is an inflammation of the stomach and small intestine, usually characterized by diarrhea, vomiting and cramps. It is sometimes referred to as the "stomach flu," although it has nothing to do with the influenza virus. "Gastro" can, however, be caused by other viruses such as rotavirus, a special problem in children, the Norwalk virus or by a variety of bacteria.

The source of Norwalk virus is fecal matter and transmission is often via water contaminated with sewage. Mollusks and other seafood that may have been exposed to sewage are another means of transmission. The virus is particularly nasty because it is highly contagious, requiring as few as ten viral particles for disease transmission. That's why one infected crew member on a ship can sicken hundreds of people. Infection comes from the ingestion of the virus from ready-to-eat food that has been handled by an infected person or by touching a surface that has been contaminated by aerosolized droplets from human emissions.

Although the virus can only reproduce once it has invaded a

cell, it can remain viable on surfaces for quite some time. How long it survives outside the body depends on a number of factors including smoothness of the surface, type of material and moisture level. Survival for several days is possible and touching the nose after touching an infected surface is a common way to spread the virus. Frequent hand washing is, of course, critical when there is an outbreak. Surfaces that are commonly touched should be washed with a solution made by diluting commercial bleach one to ten. A contact time of about ten minutes is needed to inactivate the virus. Since Norovirus is quite heat stable, it can even survive in cooked foods that are not extensively heated.

The symptoms of diarrhea and vomiting and, in about one-third of people, fever, begin anywhere from fourteen to forty-eight hours after exposure and usually last two to three days. But an infected person can still be shedding viruses for several days after symptoms have resolved. The biggest risk of an infection with Norwalk virus is dehydration, which can be severe, and in the case of the young, the old and the infirm, can be life threatening if intravenous fluids are not administered.

Norwalk virus is not the only virus that can be transmitted by food or water. Hepatitis A virus is another example, but this one has an incubation period of four to six weeks, so the source is harder to identify. It doesn't cause diarrhea or vomiting, but rather a general feeling of malaise, often accompanied by fever and nausea. The telltale sign is jaundice, whereby the skin and eyes take on a yellowish tinge. Besides human feces, mussels, clams and oysters that are harvested from sewage-contaminated water can spread the virus, as can raw fruits and vegetables irrigated with tainted water. Transmission via the fecal-oral route from person to person is possible either through skin contact, such as by shaking hands, or by handling food without proper hand washing after using the bathroom.

Questions also arise about the possible transmission of microbes by handling objects such as reusable shopping bags. These are becoming more popular as movements to eliminate disposable polyethylene bags gather steam. Reusable bags are deemed to be more friendly to the environment because they use fewer petroleum resources and do not pollute the environment. Pictures abound of plastic bags caught in trees, floating in lakes and rivers and, in some cases, literally choking wildlife. But as is often the case, a solution to a problem can introduce another problem — in this case the potential cross-contamination of food by microbes that have set up residence in reused bags.

Virtually all reusable bags tested contain bacteria, which of itself doesn't mean much because bacteria can be found everywhere. But some of the bacteria are potentially pathogenic, meaning capable of causing disease. If bags harboring bacteria from meat juices are stored in the trunk of a car for two hours, bacterial counts can increase tenfold, although this effect is temperature dependent. If the temperature rises to 53°C (127°F), possible in a trunk on a hot day, the bacterial count actually goes down.

Reusable bags can also spread bacteria in the supermarket. Shoppers usually place the bags in the baby carrier part of the shopping cart which may already have a good dose of bacteria from babies doing what babies do. Shoppers handle the bags and then may contaminate items on the shelf as well as the checkout counter. To what extent reusable bags are responsible for the spread of illness is impossible to say, but gastrointestinal ailments caused by food-borne bacteria and viruses do affect millions of people every year. Reusable shopping bags should be regularly tossed into the washing machine and then the dryer.

In one case at least, a reusable shopping bag was unfairly blamed for a Norovirus outbreak among members of a girls' soccer team in Oregon. The outbreak was indeed traced to a shopping bag that had been stored in a hotel bathroom by a team member. Its contents, sealed packages of cookies and grapes, were handled by others who then touched their face and became infected. But the virus was not brought in with the bag. One of the girls was infected and shed the virus all over the bathroom. The shopping bag just happened to be in the wrong place — any other item would have been contaminated as easily. When Norovirus-infected people do what the virus makes them do, it must be assumed that surfaces become contaminated, whether these be the faucet, the countertop, the toilet's flush handle, towels or a shopping bag that just happens to be sitting there.

Remember that food contaminated by viruses or bacteria may look and taste perfectly fine. And of course you can't see microbes. One billion of them can be frolicking on the head of a pin! It would not be a good idea to put that pin in your mouth!

WE SHARE OUR BODIES WITH BACTERIA

"I then most always saw, with great wonder, that in the said matter there were many very little living animalcules, very prettily a-moving. The biggest sort . . . had a very strong and swift motion, and shot through the water (or spittle) like a pike does through the water. The second sort . . . oft-times spun round like a top." It was with that observation in 1674 that Antonie van Leeuwenhoek etched his name in the annals of history as the "Father of Microbiology." The little "animalcules" seen cavorting in a sample of his own saliva would later be identified as bacteria. No one had noted the creatures before for the simple reason that they can only be seen through a powerful microscope. Contrary to popular belief, van Leeuwenhoek did not invent the instrument, but he did make the first microscope with magnification great enough to see bacteria.

As early as the fifth century BC, Egyptian hieroglyphs depicted simple magnifying lenses, likely the result of an accidental discovery stemming from glass making. Exactly who first made glass isn't clear, but ancient Syrians, Mesopotamians and Egyptians are the candidates. Glass is made by heating sand, soda ash and lime, all of which were available to the ancients. Soda ash is found in the residue of a wood fire and lime is calcium oxide obtained by heating limestone, which is calcium carbonate. The discovery of magnification was probably made when a piece of glass turned out to be thicker in the middle than at the ends and had a curvature to it. Such pieces were also referred to as a "burning glass" because they magnified the rays of the sun and were able to set wood on fire. The Romans made various glass containers for water and noted that looking through a globe filled with water resulted in magnification.

Curiously, though, these observations were not put to practical use until the thirteenth century in Italy when the first eyeglasses appeared. A classic painting of the era by Tommaso da Modena depicts cardinal Hugh de Provence sporting eyeglasses as he reads the scriptures.

Dutchmen Hans and Zacharias Janssen are credited with making the first microscope in 1590 by placing two lenses in a tube. Londoner Christopher White improved upon the invention and made the famous instrument that led to Robert Hooke's classic 1665 work *Micrographia*, in which he describes various objects he observed under the microscope. Hooke thought that the tiny structures he saw when looking at a sample of cork resembled the small rooms inhabited by monks and coined the word "cell," based on the Latin "*cella*" meaning "small room." Hooke's microscope, which now resides in the National Museum of Health and Medicine in Washington, was not powerful enough to see bacteria, but it did allow him to examine fossils very carefully and formulate early ideas of evolution.

Then along came van Leeuwenhoek, who had no training as a scientist but had developed an interest in lenses working in a dry goods store where magnifying glasses were used to count the threads in cloth. Van Leeuwenhoek thought he could improve upon the magnifiers used at the time and developed new methods for grinding and polishing tiny lenses of great curvature that magnified over two-hundred-fold. He had an innate curiosity that led him to examine almost anything that could be placed under his lens, and in a series of letters to the Royal Society of England describes the teeming life in a drop of water, red "corpuscles" in blood and sperm cells in his own semen, which he stresses were acquired not by sinfully defiling himself but as a natural consequence of conjugal coitus.

Although van Leeuwenhoek unveiled the microbial universe,

he had no idea of the stunning number of "animalcules" that populated that universe or the role they play in human lives. Incredibly, the total mass of bacteria on earth exceeds the mass of all plants and animals combined! Our bodies harbor about ten times as many bacterial cells as human cells, so if you go by sheer numbers, we are actually more bacteria than human! And we are learning more and more about how changes in the population of the thousands of different types of bacteria that inhabit our body, particularly our digestive tract, can affect our health. These bacteria usually keep each other in check as they compete for the same food supply, but either an infection or the use of antibiotics can upset this balance.

To pick one interesting example, consider the case of an Australian patient with Parkinson's disease who suffered from protracted constipation. His physician thought that perhaps the constipation was caused by an infection and decided to try treatment with an antibiotic. To everyone's surprise, the Parkinson's symptoms abated!

Is it possible that Parkinson's disease in some cases is caused by gut bacteria entering the vagus nerve that connects the digestive tract to the brain? Sounds far-fetched, but the suggestion that ulcers are caused by the *Helicobacter pylori* bacterium or that *Clostridium difficile* infections can be treated with fecal transplants or with pills loaded with microbes derived from human feces were also initially greeted with derision. Not only do fecal bacteria treatments work, there have even been cases where such treatment resulted in improvements in other conditions the patient was suffering from such as multiple sclerosis, rheumatoid arthritis, chronic fatigue and Parkinson's disease. Microbial populations in the gut have even been linked with obesity. Overweight people may have higher populations of bacteria that release nutrients from food that otherwise might pass through the system undigested. This notion is backed by studies in mice. Transferring bacteria from the guts of obese mice into lean ones causes the lean mice to put on weight.

Surely van Leeuwenhoek would be surprised by all the activities of his "animalcules" uncovered by modern microbiologists and undoubtedly would be keen to investigate them further. As he stated in a letter to the Royal Society in 1716, "My work . . . was not pursued in order to gain the praise I now enjoy, but chiefly from a craving after knowledge." It is such cravings that satisfy our appetite for scientific breakthroughs. And who knows, maybe one of those breakthroughs will involve improving our health by dining on some of Van Leeuwenhoek's little "animalcules."

GETTING STEINACHED

The Great War was over, cars were multiplying on the streets, radios were crackling in living rooms, plastics were hitting the market and theaters were attracting people with new-fangled moving pictures. Science and technology were roaring ahead. It was, after all, the "Roaring Twenties." But in Vienna, there was another kind of roar. It was emanating from thousands of older men who claimed to have regained their virility through what seemed to be a stunning advance in medicine. They had been "Steinached"! The men had undergone a twenty-minute procedure introduced by Dr. Eugen Steinach in which one of their seminal ducts was tied off. In other words, the men underwent a partial vasectomy. The goal wasn't prevention of pregnancies, it was rejuvenation!

Steinach's work was stimulated by French physiologist Charles-Édouard Brown-Séquard's seminal lecture delivered to members of La Société de Biologie in 1899 in which he described having injected himself with filtered extracts from the crushed testicles of young dogs and guinea pigs to regain the vigor and intellectual stamina of his youth. The professor had also tested himself with a dynamometer, a device that measures mechanical force, and found that his muscle strength had also been renewed. He capped off the lecture by telling his rapt audience that just hours earlier he had passed the final test of his experiment by "paying a visit" to his young wife.

The scientific community, however, did not buy Brown-Séquard's claim that the key to rejuvenation was injection of minced gonads. The prestigious *Boston Medical and Surgical Journal* opined that "the sooner the general public and especially septuagenarian readers of the latest sensation understand that for the physically used up and worn out there is no secret of rejuvenation, no elixir of youth, the better." Biology professor Eugen Steinach, however, thought that Brown-Séquard's work with gonads was worth pursuing and turned to transplanting the testes of a male guinea pig into a female. She then exhibited mounting behavior characteristic of a male! Steinach concluded that the gland's secretions were responsible for sexuality and even theorized that homosexuality in men could be treated by transplanting a testicle from a "normal" man into a recipient in need of "remasculinization." Thankfully that idea didn't fly, but surgeon Serge Voronoff's notion of grafting monkey gland tissue onto the testicles of aging men did!

While serving as physician to the king of Egypt, Voronoff had noted that the court eunuchs were often sickly and seemed to age very quickly. The testes, he concluded, played an important role in maintaining vigor, and that "possession of active genital glands was the best possible assurance for a long life." In 1918, he believed he made his point when he restored an aging ram's youthful vitality by transplanting the testes of a young lamb.

Voronoff upped the ante by transplanting the testes of executed criminals into aging men rich enough to pay for the procedure. But demand soon outstripped supply, and since few young men were willing to part with their precious parts even for rich compensation, Voronoff came up with an alternate scheme. He would transplant bits of chimpanzee and monkey testes onto the genitals of elderly men. Eventually more than a thousand men underwent the monkey-gland treatment at the hands of doctors around the world, with the requisite material often being supplied by a monkey farm Voronoff set up on the Italian Riviera.

Steinach bought into Voronoff's idea, but thought that the benefits ascribed to transplants could be achieved by an alternate procedure. Damming the seminal canal would stimulate the testes to produce more male hormones! At the time, researchers had determined that that there were two types of tissues in testicles. Seminal tubules produced spermatozoa, but between the tubules there were are also Leydig cells that released sex hormones. Steinach's idea was that the two types of tissues compete for nourishment, and that stifling the sperm-producing tissues would boost the production of the sex hormones.

In his book, *Sex and Life*, Steinach described how his patients "changed from feeble, parched, dribbling drones, to men of vigorous bloom who threw away their glasses, shaved twice a day, dragged loads up to 220 pounds, and even indulged in such youthful follies as buying land in Florida." He believed in his procedure so strongly that he "thrice reactivated himself." It isn't clear what he meant by thrice, because once the duct is tied off, it's tied off. Whatever improvement Steinach and his patients felt was likely due to wishful thinking, because as we now know, vasectomies do not boost hormonal output by the testes.

Steinach had testimonials galore, including from some very famous people such as Sigmund Freud, who underwent the procedure when he was sixty-seven years old, hoping to improve his "sexuality, his general condition and his capacity for work." William Butler Yeats, the famed writer, was Steinached when he was sixty-nine. "It revived my creative power," wrote Yeats in 1937. Apparently in more than one way. The doctor who performed the snip invited a woman half Yeats's age to dinner with the aim of allowing the writer to make a connection and test out his newly embellished virility. It seems the outcome was successful, with Yeats publicly reporting on his "second puberty," leading to the Dublin press nicknaming him the "gland old man."

While it is now clear that Brown-Séquard, Voronoff and Steinach promoted procedures that did not have the claimed efficacy, they did lay the foundations for further research that resulted in the isolation of testosterone, the male sex hormone. In 1927, University of Chicago Chemistry professor Fred Koch isolated 20 milligrams of a substance from 20 kilos of bull testes that remasculinized castrated roosters, pigs and rats. By 1935, Schering's Adolf Butenandt had worked out the molecular structure of testosterone, the active compound in the bull testes, which allowed him to come up with a chemical synthesis from cholesterol. Today, testosterone and various derivatives are prescribed to men with low blood levels. They often claim to experience the effects that were thirsted for by men who subjected their privates to the scalpels wielded by Drs. Steinach and Voronoff in the roaring twenties.

BEETHOVEN'S POULTICE AND MOZART'S PORK CHOPS

Both Beethoven and Mozart were spectacular talents whose music will live forever. That's a fact. But when it comes to the composers' deaths, facts give way to theories — some reasonable, some outlandish. Posthumous investigation in the face of a lack of sufficient evidence amounts to no more than conjecture, so why even bother? Because the theories do offer an opportunity to discuss some interesting science.

Ludwig van Beethoven died in 1827. That much we know. But what killed him? Many experts and pseudo-experts have attempted to answer that question. Some even suggest that the question should actually be "who killed him?" Beethoven's progressive deafness is well documented, but that was not his only ailment. From a young age, the composer complained of abdominal pain, headaches and diarrhea, and as the years passed, his days were plagued with fits of aggressive behavior, impulsiveness and depression. When he passed away at age fifty-six, an autopsy revealed liver and kidney disease. What could have caused these problems?

Samples of Beethoven's hair and bones would eventually become the focal point for the numerous discussions and articles about his death. In 2005, analysis of the hair samples and skull fragments indicated a higher than normal level of lead, generating widespread speculation about the possibility of lead poisoning.

Could it have been caused by lead in waters from the spa he frequented to alleviate his ailments? Perhaps drinking from lead goblets? Could it have been the wine he was so fond of? Or was the culprit a lead-laced poultice his physician used after withdrawing excessive fluid from Beethoven's abdomen? That's the theory forwarded by Viennese pathologist Dr. Christian Reiter.

He claims that Beethoven was primed for a calamity because he already had accumulated high levels of lead from his favorite wine and the surgical procedure put him over the top.

Although thin sheets of lead were used at the time to keep poultices in place, there is no evidence that the doctor used this technique. In fact, his own account describes that the puncture wounds were being kept "meticulously dry in order to avoid infection." But the wine story has a ring of truth to it. Beethoven was not an alcoholic, but he did love wine. His last words on his deathbed supposedly were "pity, pity, too late" after being told of a gift of twelve bottles of wine. And apparently his preference was for some cheap Hungarian wine, probably because of its sweetness.

At the time, many such wines were adulterated with lead acetate to improve the flavor. Beethoven's friends spoke of him drinking a bottle of wine with each meal, and his love of wine was well expressed in his classic quote: "Music is a higher revelation than all wisdom and philosophy, it is the wine of a new procreation, and I am Bacchus who presses out this glorious wine for men and makes them drunk with the spirit." But it was his imbibing in real spirit that may have done him in. And I say "may," because the most recent analysis of bone fragments and hair using more sophisticated equipment does not show an unusually high level of lead. So the possibility of lead poisoning is interesting but doesn't hold a whole lot of weight.

The prospect of poisoning has also been raised in the death of Mozart at age thirty-five after a brief illness. In 1791, the most famous child prodigy of all time suddenly developed a high fever, sweats, nausea, diarrhea, a rash and severe pain in the hands and legs. His body swelled terribly and emitted a horrific odor. Russian writer Aleksandr Pushkin's 1830 play *Mozart and Salieri*, raised the possibility that Salieri, who was court composer at the time, poisoned his young rival in a fit of jealousy. Another candidate for poisoning Mozart, who was known to be a notorious womanizer, was Franz Hofdemel, whose wife had apparently reciprocated the composer's amorous advances. And as one might expect, with prejudices running wild at the time, blame even fell on the Freemasons, Catholics and of course the Jews. Arsenic was the most common poison at the time, but Mozart's symptoms do not match those of arsenic poisoning, with burning of the throat and difficulty in swallowing being notably absent. Neither do the symptoms mesh with mercury poisoning, which has also been suggested. The supposition is that Mozart suffered from syphilis, and accidentally self-administered a toxic dose of mercury chloride, a common medication at the time.

There is no shortage of conjecture about Mozart's illness and death. Kidney disease brought on by a streptococcal infection is a realistic possibility, especially when considering that there was a minor epidemic of swelling-related deaths in Vienna at the time. Rheumatic fever, infective endocarditis, vitamin D deficiency, and a rare autoimmune disease called Henoch-Schonlein purpura have also been suggested, as well as the composer's love of pork chops. Trichinosis, caused by a parasitic worm that infects pork, can cause symptoms similar to what Mozart experienced, although the progress of the disease would be different from what observers described.

In the absence of a body, one of course can do no more than make guesses. And there is no body. But there may be a skull. According to custom at the time, Mozart's body was sewn in a linen sack, placed in a communal grave and doused with quicklime to hasten decomposition. This allowed the graves to be opened years later and be used again after dispersal of the remains. It seems the worker who opened the grave was the same as had buried Mozart and remembered where the head had been. He snitched the skull, which is now on display at the Mozarteum in Salzberg.

French anthropologist Pierre-François Puech claims that the skull is indeed Mozart's, basing his opinion on a developmental abnormality in the skull that leads to a forehead just as depicted in paintings of the composer. Furthermore, the skull shows a fracture that Puech says was caused by Mozart falling and led to a brain contusion that caused his death. His conclusion has been disputed both on grounds of Mozart's symptoms and the fact that Mozart's physician had recorded that his patient had only seven teeth. The skull has eleven. Mozart was apparently a lot better at writing music than brushing teeth.

THE FOUNTAIN OF YOUTH AND ALLIGATORS

Funny the things one remembers. Like "Don's Fountain of Youth," a short cartoon I saw some time back in the 1960s. "Don" was Donald Duck, and the story was all about taking his nephews on a Florida vacation. The kids are more interested in reading comics than the sights that Donald is pointing out, at least until they chance upon a pond with a sign "Mistaken for the Fountain of Youth by Ponce de León 1512." Donald decides to have a little fun with his nephews and removes the "mistaken for" part of the sign. He wades into the water and pretends to have turned into an egg that he found in a nearby nest.

I was intrigued by that cartoon for a couple of reasons. I was sort of a history buff and wondered whether the reference to Ponce de León was real or fictional. And the "Fountain of Youth" caught my attention because I had grown up with a special fondness for perhaps the most famous Hungarian musical, *János Vitéz* (*John the Valiant*), based on a poem by the celebrated poet Sándor Petőfi in which the "Spring of Life," capable of conferring immortality, plays a vital role.

Checking the *Encyclopedia Britannica* in those pre-Google days revealed that Ponce de León was indeed a real Spanish explorer who gave Florida its name upon seeing a landscape filled with flowers. Stories about his search for a Fountain of Youth, however, did not emerge until after his death and appear to be mythical. But tales of waters that restore youth have been with us for thousands of years. Alexander the Great supposedly looked for a restorative spring, and *The Fountain of Youth*, a famous 1546 painting by Lucas Cranach the Elder, depicts aging people entering a pool on one side and emerging as youths on the other.

These days the search for the fountain of youth is more likely to lead us to a bottle than a spring or lake. There are waters that are oxygenated, hydrogenated, alkalized, ionized, clustered, frequency harmonized, catalyst altered, photonically enhanced, plasma activated, vibrationally charged, chi energized or DNA encoded. Their ads flood us with scientific-sounding lingo and promises of health and rejuvenation. But the science is all wet and drips with crackpot notions. If a fountain of youth is to be found, it may indeed be found in a bottle, but it won't be filled with "penta," "double helix" or "negative field activated" water. It will likely contain a pharmaceutical product of some sort. Maybe something like metformin.

Metformin is the most widely used diabetes drug in the world. It decreases glucose production in the liver, increases insulin sensitivity and enhances the uptake of glucose from the bloodstream by cells. All of this amounts to reducing the amount of glucose in the blood, the hallmark of diabetes control. But why the connection to aging? Because a universal sign of aging is the development of insulin resistance, a condition in which cells progressively fail to respond to insulin, the cell's gatekeeper for the entry of glucose. As insulin resistance develops, blood levels of glucose rise and diabetes eventually sets in. Since diabetics are at an increased risk for cancer, heart disease and inflammatory conditions, it comes as no surprise that metformin has been associated with a decreased risk of these ailments. Given that such illnesses are normally age related, researchers began to wonder whether aging was actually a sort of pre-diabetic condition that could be retarded with the appropriate use of metformin.

When scientists are confronted by such a premise, they tend to turn to mice. Mice may not be men, but they can be given doses of a drug, their diet can be controlled and their health status and lifespan can be readily determined. When researchers at the National Institutes of Health in the U.S. treated mice with metformin at a dose of 0.1 percent by weight of their total food intake, the animals' mean lifespan was extended by about 6 percent! Interestingly, even though the treated mice had consumed more calories, they weighed less than the controls, likely because of an increased use of fat for energy. Confirming the notion that the difference between a drug's ability to cure or to kill lies in the dose, a tenfold increase in the amount of metformin given resulted in a 14 percent shorter lifespan.

The mechanism by which this drug extends life isn't clear but some clues are emerging from research with *Caenorhabditis elegans*, a tiny worm that is commonly used to study aging because it has a lifespan of only three weeks. Also, their aging process can be followed visually because the little creatures get smaller, wrinkle up and move more slowly. That is, unless they are treated with metformin. Then they lead longer, healthier lives. And it seems that effect has to do with metformin's ability to generate a small dose of reactive oxygen species (ROS). These are generally a normal product of metabolism but because of their free radical character are highly reactive and can cause cellular damage. Indeed such oxidative stress has been associated with aging, and theories abound about the use of antioxidants to counter the process. But now Belgian researchers have cast a shadow on this widely accepted theory by demonstrating that metformin produces a small dose of ROS that actually increases the robustness and longevity of a cell. If this effect can be confirmed, it would imply that antioxidants could actually negate the benefits of metformin.

Of course we are neither mice nor worms, so we will have to wait for more evidence about metformin's benefits before jumping on the anti-aging bandwagon. Many wagons carrying the likes of human growth hormone, dehydroepiandrosterone

(DHEA), shark cartilage extracts and megavitamin supplements have ground to a halt after a promising start. Metformin, however, may yet hatch into an effective life-extension treatment. But let me take you back to Donald for a moment. The egg he picked up to trick his nephews into believing in the anti-aging effect of the Fountain of Youth turned out to be an alligator egg. And that resulted in an unforeseen outcome, namely Donald being chased by a pretty angry alligator mom. That chase, though, had a real anti-aging effect. Exercise really does turn back the clock. Just look at Donald — he hasn't aged a bit. He turned eighty in 2014.

STRETCHING THE TRUTH

OF MICE AND MEN

Let's face it, running on a treadmill isn't one of life's most exciting activities, but it does provide time to contemplate life and think about what is likely to extend it. There's plenty of evidence that exercise will, which is why one plods away on the treadmill. But should one gear up for short bursts of high-intensity exercise or scamper along at a slower pace for a longer time? The scientific literature is ambivalent on the issue, but it is one that I follow closely because I am sort of addicted to the treadmill. That's why a *New York Times* blog with the headline "For Fitness, Push Yourself" accompanied by a photo of competitive runners obviously at full tilt got my attention. "Intense exercise changes the body and muscles at a molecular level in ways that milder physical activity doesn't match, according to an enlightening new study," the article began. Was there finally an answer to the exercise conundrum?

The study was enlightening all right, if you are a mouse. This is not a criticism of the research, which was carried out by a very reputable group at the Scripps Research Institute in Florida. But it is a criticism of the interpretation of the study not only by the *New York Times* blog, but by many other media reports. The study states, "To realize the greatest benefits from workouts, we probably need to push ourselves." There were also quotes from one of the researchers involved in the study about "no pain, no gain." Coming to such a conclusion based on a study involving specially bred mice scuttling on a treadmill is way too adventurous.

The study's basic goal was to examine how the hormones adrenaline and noradrenaline affect muscle structure. These hormones are released under stressful conditions and are known to prime muscles for "flight or fight." Since intense exercise is also known to release these chemicals, it is reasonable to explore its potential to increase muscle strength. The effects of the stress hormones are thought to be manifested through the activation of a specific protein termed CRTC2, present in mice as well as in people. The Scripps researchers therefore bred mice that were genetically programmed to produce more of this protein, put them on a program of strenuous treadmill exercise and found that they developed larger muscles and were more efficient at releasing fat for use as fuel than control animals. Interesting, but genetically modified mice are a long way from humans, and the study does not justify giving any sort of advice to people.

The researchers also talk about "searching for molecular therapeutics that will activate the CRTC2 protein so that "even an average exercise routine could potentially be enhanced and made more beneficial." Sounds like an attractive research project, but I suspect it won't be long before an inventive marketer puts the cart before the horse and starts promoting some sort of "CRTC2 enhancer."

In the anti-aging business, making more of reputable science than is warranted is par for the course. Consider these headlines: "Cocoa Extract Highly Effective in Protecting Against

Alzheimer's Disease, Says New Study" or "Worried About Alzheimer's? Go on a Chocolate Binge, Study Says." Well, no. The study doesn't say anything like that. The grossly exuberant headlines were prompted by a paper published in the *Journal of Alzheimer's Disease* entitled "Cocoa extracts Reduce Oligomerization of Amyloid-beta: Implications for Cognitive Improvement in Alzheimer's Disease."

Did the researchers from the Mount Sinai School of Medicine in New York carry out experiments with cocoa on Alzheimer's patients? No. Did they feed cocoa to animals? No. What they did was study the effects of a specific type of cocoa extract on the activity of nerve cells in mouse brain tissue dosed with synthetic compounds thought to model Alzheimer's disease.

One of the hallmarks of Alzheimer's disease, which affects an estimated 36 million people worldwide and is expected to double by 2030, is the deposition of a protein known as amyloid-beta between nerve cells. This virtually gums up the workings of the brain by preventing neurotransmitters, the chemicals nerve cells use to communicate with each other, from crossing the synapse, the gap between nerve cells. Since amyloid proteins are formed from smaller fragments called peptides, any interference with the ability of peptides to aggregate into the troublesome proteins is worthy of investigation.

Flavanols are a class of compounds found in cocoa that have been proposed as candidates for interfering with the formation of the amyloid proteins. The Mount Sinai researchers decided to use an unfermented, lightly processed cocoa known as "lavado" in their investigation because of its high flavanol content. Most commercial cocoa is "Dutched" and has undergone alkali treatment to reduce bitterness, a treatment that also significantly reduces flavanol content. As far as chocolates go, their flavanol content is minimal.

The experiment that generated all the publicity consisted of bathing brain slices from mice specially bred to be prone to Alzheimer's disease in solutions of the amyloid precursor peptides mixed with different cocoa extracts. When the nerve cells in these tissues were electrically stimulated, transmission of information between them was enhanced with Lavado cocoa extracts.

While this is interesting research, it cannot be used to draw any conclusion about people consuming cocoa. There is no way to know how the amount of the cocoa extracts used in these experiments relate to amounts of flavanol that may make it to the brain from eating chocolate or drinking cocoa. And mouse brain slices in a lab are a long way from a functioning human brain. Although maybe not so far from the human brains that clutter the media implications for human health based on preliminary laboratory or animal experiments.

Needless to say, I won't go out searching for Lavado cocoa, at least not until a proper randomized trial in humans shows a benefit. And as far as the treadmill goes, I have no idea what "intense" mouse exercise means in human terms, but on looking into the issue I did come across a scientific paper that added some pep to my treadmilling. The title was "Physical Exercise Protects Against Alzheimer's Disease." I won't be shouting about it from rooftops though: the study was on mice genetically modified to develop the disease.

ACUPRESSURE MATTRESS DOESN'T NAIL IT

Some time ago, I spent hours hammering hundreds of long nails through a plank of plywood. Wasn't easy. The nails had to be carefully spaced, about a centimeter apart, protruding exactly the same distance from the wood. Any deviation would have made it quite uncomfortable to lie on my new bed of nails! The point, as it were, was to demonstrate to students that you did not have to be a Hindu mystic to lie on a bed of nails. There was absolutely nothing paranormal about accomplishing the feat. It was simply a question of physics. As long as the weight was distributed over enough nails, there was no worry about skin penetration.

While this was a neat demonstration, I can't say it was particularly relaxing. That's why I was taken a little aback when I came across what amounted to a bed of nails being sold in a health food store with claims of promoting relaxation and stress reduction. Not only that, it promised to energize, improve sleep, reduce pain, increase circulation and, within five minutes, provide a fresh glow and face-lift effect. Meet Spoonk, the "acupressure massage eco mat"!

Spoonk is a flexible plastic mat that, instead of nails, features 6,210 sharp plastic stimulation points. The odd name is a whimsical version of "spunk," a word made up by Pippi Longstocking, the fictional heroine created by Swedish children's writer Astrid Lindgren. While Pippi attached no meaning to the word, it became associated with strength, energy and a love of life, all characteristics Pippi possessed. The apparent message is that Spoonk can bestow these very properties. The rationale is based on the concept that the body is permeated with channels called meridians through which a sort of life energy, often referred to as chi, flows. Any blockage of the flow of chi means bad news.

According to Traditional Chinese Medicine, these blockages can be cleared either by the appropriate application of needles, as in acupuncture, or with physical pressure. Such "acupressure" is said to be the principle behind Spoonk. The little spikes are designed to produce some sort of a shotgun effect, clearing all the possible meridian blockages. Anatomical science, however, cannot detect any sort of meridian and no measurable chi force exists. Of course that does not mean that acupressure cannot work by some other means.

My first encounter with acupressure was back in the 1960s in an introductory psychology course at McGill, although the term was never mentioned and there was no talk of any chi. I was lucky enough to attend a lecture by Professor Ronald Melzack, one of the world's premier experts on pain and developer of the McGill Pain Questionnaire, used by cancer clinics around the globe. Frankly, I don't think I really understood his "gate theory," and still don't, but I know it has something to do with the spinal cord either blocking pain signals or allowing them to pass to the brain, depending on whether the signal travels via small nerves or through larger nerve fibers. Somehow by applying appropriate pressure in certain spots, the "gate" that allows a pain message to pass to the brain can be blocked. It all sounds very theoretical, but Dr. Melzack's practical example made an impact. If you have a toothache, he advised, just rub your hand between the thumb and forefinger with an ice cube. That sounded pretty odd, but Professor Melzack wasn't talking out of his hat — he actually quoted a clinical trial that had demonstrated success.

Of course rubbing a hand with an ice cube for relief of a toothache is a far cry from relaxing the body by lying on a plastic bed of spikes. Or energizing it. Actually, it isn't clear to me how you can be both energized and relaxed at the same time, but never mind that. The pertinent question is whether there is any evidence that the mat provides benefit. I can't find any trials that have put Spoonk to a test, but I did turn up some sketchy data about a handmade mat with some 1500 stainless steel office pins, invented by a Russian layman, Ivan Kuznetsov, back in the 1980s. He figured that by poking the body everywhere, he would be hitting some of the right acupuncture points, and no harm would be done by any that were off target. The mat was eventually sold in pharmacies and spurred a television documentary that detailed successes in alleviating pain. No studies, however, were published in the scientific literature.

An American version of Kuznetsov's mat was marketed for a while under the name Panacea. In one study, albeit not a methodologically impressive one, of 200 users, 98 percent reported pain relief, 96 percent reported relaxation, 94 percent reported improvement in the quality of sleep and 81 percent reported an increase in energy level. Approximately half of the subjects with allergy problems reported relief of their symptoms. The fly in the ointment here is that subjects commonly report such benefits no matter what kind of treatment they are offered. Similar benefits are claimed by people who rub their body with snail slime, drink oxygenated water, sport plastic bracelets with "therapeutic" holograms or bask in the reflected light of the Arizona moon.

While there are no compelling studies about acupressure mats, there have been a surprising number of studies on various other forms of acupressure. In fact, some forty-three studies have investigated stimulating various body parts in order to manage pain, breathing problems, fatigue, insomnia and pregnancy-related discomfort or chemotherapy-induced nausea and vomiting. While some studies have shown benefit, a systematic review published in the *Journal of Pain and Symptom Management* concluded, "Our review of clinical trials from the past decade did not provide rigorous support for the efficacy of acupressure for symptom management."

Spoonk asks us to "imagine a world free from stress, anxiety, depression, stiffness and pain." The company's stated mission is "to contribute to that imaginary world with a simple and effective product." I can go with the "imaginary," but the use of the term "effective" would have left me somewhat skeptical had I not noted the logo of *The Dr. Oz Show* on Spoonk's package. Surely Dr. Oz would not lead us astray! What a stressful thought. So I dug out my old bed of nails. After all, according to Spoonk, stimulating all those acupressure points is a great stress-buster. Didn't work. Tore my pants.

POPEYE'S FOLLY

Mea culpa. I plead guilty to the crime I often accuse others of committing, namely not checking facts properly! Curiously, I would not have discovered my error had I not been doing some proper fact-checking about claims that a nutritional supplement derived from the root of the maca plant can increase libido and alleviate menopausal problems. While looking into this, I came across an ad from Popeye's, a sports nutrition company that sells maca extract. And then as I searched the web for info about this supplier, I came upon an article by criminologist Dr. Mike Sutton: "Spinach, Iron and Popeye—Ironic lessons from biochemistry and history on the importance of healthy eating, healthy skepticism and adequate citation."

I was immediately intrigued, given that I had once written an article in which I talked about how cartoonist Elzie Segar attributed the strength Popeye needed to rescue his beloved Olive Oyl from the clutches of the dastardly Bluto to the high iron content of spinach. I described that Segar had actually overstated the vegetable's iron content because of a mistake in the scientific literature. Apparently a German researcher in the late nineteenth century had placed a decimal point in the wrong place, ascribing to spinach a tenfold greater iron content than was actually present. I proceeded to relate how Popeye's fondness for spinach led to a huge increase in sales that resulted in statues honoring the sailor man being erected in spinach-growing areas to commemorate his prodigious consumption of the vegetable. And I pointed out that it all was because of an error in the placement of a decimal point! Alas, as it turns out, the error was mine. There was never any decimal point mistake, Popeye was not responsible for a dramatic increase in spinach sales and, most strikingly, Elzie Segar had never claimed that Popeye's strength was due to iron in spinach!

Where did I get my story? I had done my research, just not well enough. My first stop had been an article published in 1977 by Professor Arnold E. Bender, a man who was no slouch when it came to nutritional research. Bender was Head of Nutrition and Dietetics at Queen Elizabeth College in England, had authored more than 150 research publications and 14 texts. His book *Health or Hoax: The Truth About Health Foods and Diets*, is a classic. In his article, Bender described the determination of the iron content of spinach by Dr. E. von Wolff in 1870 and how a subsequent analysis in 1937 by Professor Schupan had found that spinach contained no more iron than any other leafy vegetable. In fact, only one-tenth the amount von Wolff reported! "The fame of spinach appears to have been based on a misplaced decimal point," Bender concluded, as it turns out, without any evidence. Given the author's stellar reputation, I saw no need to delve further into the matter especially, when a 1981 article in the *British Medical Journal* by hematologist Dr. T. J. Hamblin apparently confirmed the story.

Hamblin's theme was how "frauds, hoaxes, fakes and widely popularized mistakes run through the history of science and medicine." He explained that linking Popeye's superhuman strength to spinach was due to a decimal point error, and how because of this error Popeye had single-handedly raised the consumption of spinach by 33 percent. He concluded that as far as iron intake went, Popeye would have been better off chewing on the cans the spinach came in. That, he may actually have been right about. Ironically, as Dr. Sutton's meticulous research shows, Hamblin was perpetuating both the myth of the decimal point error and the belief that Popeye guzzled spinach for its iron content.

Sutton did a remarkable job tracing the confused story of Popeye's link with spinach. He did what all good scientists should do: he checked the facts, if possible by going to the original source. While there was no decimal point error, the German chemists who determined the iron content of spinach in the late 1800s did make some mistakes. There may have been contamination of samples with iron from laboratory equipment, and there was also confusion about whether the iron they found referred to that in fresh spinach or the dried variety. A cup of dried spinach would have much more iron given that fresh spinach has a high water content. The fact though, is that while early analysis may have overstated the iron content of spinach without any decimal point error, the correct value was already well known before Popeye was ever conceived.

So is spinach a good source of iron? Yes and no. A cup of cooked spinach contains about 6.5 mgs of iron, which is a fair amount considering that an average person needs about 8 mgs a day. Premenopausal and pregnant women need 18 and 27 mgs respectively. A cup of raw spinach has less than 1 mg because of the high water content. But there's another issue. Spinach is high in oxalic acid, which inhibits iron absorption. Basically, spinach is not a great source of iron. And as far as iron providing extra energy goes, that would only be the case if weakness were due to iron deficiency anemia. Popeye, being a sailor, is unlikely to have suffered from such a deficiency, given that seafood is an excellent source of heme iron, the most readily absorbed form.

Now for the real crime that Sutton uncovered. Elzie Segar never made any mention of iron in connection with spinach! His first reference to spinach was in a 1932 strip that shows Popeye munching away and declaring that "spinach is full of vitamin A an' tha's what makes hoomans strong and helty." Segar clearly attributed spinach's value to vitamin A, not iron.

He erred a little here since spinach contains no vitamin A, but does indeed contain a good dose of beta-carotene that the body converts into vitamin A. (Interestingly, vitamin A helps mobilize iron from its storage sites, so a deficiency of vitamin A limits the body's ability to use stored iron, resulting in an "apparent" iron deficiency.)

In summary, there was never a decimal point error in the determination of the amount of iron in spinach, the vegetable is not a particularly great source of iron, and Popeye never claimed his strength came from iron in spinach. The moral of the story? Even when it comes to apparently trusted authorities, check the facts! And as far as the claims about maca go, I'm going to do some very careful checking.

BALANCING HYPE AND SCIENCE

Some memories remain indelibly etched in one's mind. Like cheering in the Montreal Forum during the 1976 Olympics as Romanian gymnast Nadia Comăneci performed miracles on the balance beam and uneven bars. The total of seven scores of ten she would eventually receive allowed her to rival Count Dracula as Romania's most famous citizen. It also catapulted Nadia into several careers, including being a spokesperson for a line of cosmetics produced by the Gerovital Company. CosmeSilk Sericin Q Complex promises to preserve youth with sericin, "a unique biopolymer with a unique structure leading to unique performance." A string of uniquely meaningless terms.

The name Gerovital undoubtedly rings a bell with European immigrants, particularly Romanians. It was back in the 1950s that Romanian gerontologist Dr. Ana Aslan introduced a potion that became famous around the world as a "fountain of youth in an ampoule." Aslan was a good friend of Nicolae Ceauşescu, the country's notorious dictator, who was keen to present a youthful and vigorous image of himself and supposedly charged Aslan with devising a remedy to turn back the clock.

Dr. Aslan, who passed away in 1988 at the very respectable age of 91, got on the track of Gerovital after hearing accounts from physicians about alleviation of arthritis symptoms and improved skin elasticity in patients who had been administered the local anesthetic procaine hydrochloride. Aslan herself carried out trials that she claimed showed the drug increased longevity in rodents. Although others were unable to duplicate these results, Gerovital managed to develop a "jet-set" aura, apparently snaring such celebrities as Marlene Dietrich, Kirk

Douglas, JFK and Nikita Khrushchev. Procaine may well have been the only thing the latter two ever had in common.

Much of the satisfaction with Gerovital was undoubtedly due to the placebo effect, but it seems the drug may have some pharmaceutical properties other than inducing anesthesia. Researchers agree that procaine hydrochloride is a weak monoamine oxidase inhibitor. In other words, it acts as a mild antidepressant, which would appear to explain the feeling of well-being claimed by its proponents.

One of the breakdown products of Gerovital in the body is diethylaminoethanol, a compound that has antioxidant properties. Such substances may indeed reduce damage to tissues caused by free radicals, but procaine is not innocuous, sometimes causing allergic reactions and migraines. There is insufficient evidence to warrant the use of Gerovital to counter the aging process, and its sale in North America is illegal. The company has, however, come up with various cosmetics that ride the coattails of Gerovital's dubious fame and promise a range of anti-aging effects.

Gerovital Anti-aging Super Enzyme cream has more than fifty ingredients including superoxide dismutase (SOD). This enzyme is found in human cells, where it plays a vital role in neutralizing superoxide, a potentially cell-damaging free radical generated by normal metabolic processes. SOD may indeed prevent skin damage when it is synthesized inside cells, but there is no evidence that it can be absorbed to any significant extent when applied topically. Given that it is third from the end on the list of ingredients, SOD is unlikely to contribute anything other than an opportunity for advertisers to tout the wonders of "bio-mimetic" ingredients and "long term anti-aging effects."

The cosmetics industry has often been castigated for such "inventive" marketing, but the industry also features some very inventive science. Strangely though, there is an interesting wrinkle here. Substantiating a claim of skin rejuvenation would require a demonstration of structural changes in the skin and a permanent elimination of wrinkles. But that would also mean the product would be classified as a drug rather than a cosmetic and would therefore require a prescription. This is the case for creams containing retinoic acid, a compound that has been shown in properly conducted scientific trials to improve the appearance of sun-damaged skin and to stimulate the growth of collagen, the protein responsible for the skin's elasticity and firmness. The challenge then for cosmetic manufacturers is to develop products that can be scientifically shown to improve the appearance of the skin without causing significant changes in its structure. A temporary ironing out of the wrinkles, as it were.

Special instruments originally designed to test the smoothness of race car tracks are now available to measure the depth of wrinkles. The fact is that all moisturizers reduce wrinkle depth to some degree by puffing up the skin as they are absorbed, but cosmetic chemists are in a constant search for ingredients that enhance this effect. Palmitoyl pentapeptide, acetyl hexapeptide-3, hyaluronic acid, furfuryladenine and a host of other compounds all claim to erase fine lines and mask blemishes. And companies do provide evidence to back up the claims with photomicrographs of skin cells and close-up pictures of improved "crow's feet." Academically interesting to be sure, but the question is whether an objective observer will note an improvement. Mirrors don't lie, but the reflection seems to reflect the amount of money spent on a product.

With the current concern about chemicals, producers are looking for "natural" alternatives. Stemlastin is described as "an extract of a particular red algae found living in extreme conditions in volcanoes in Indonesia produced in a natural, eco-friendly way, using a photo bio-reactor that delivers a special intra-cellular composition of extremolytes like mineral nutrients, amino acids and algae polyphenols." I have no idea why some algae's ability to survive in volcanoes has anything to do with wrinkles, but the impressive-sounding hype could well erupt into hot sales.

Neither is it clear why the gummy coating on the fibres produced by the silkworm should "restore skin tonicity and firmness" as claimed by Nadia Comăneci Skincare's Blossom concoction. I remain unimpressed by the claim that "the Empresses of China and Japan used it for centuries to stretch their youth." I think what is being stretched is the truth. I have no doubt about the product's moisturizing effect, something that is in the realm of any protein preparation. But I do wish that Nadia's talent on the balance beam had translated to a better ability to balance hype and science in her line of cosmetics. Still, because of the thrill she gave me and others in 1976, I'll forgive this little wrinkle in her career.

HOMEOPATHY BUGS ME, BUT NOT THE BUGS

"It's natural medicine." "Herbal remedies." "It's like . . . acupuncture." "No chemicals." "Energy healing." "Yogurt." Such were the common responses when we asked random people, including physicians, about their thoughts on homeopathy. All wrong. Homeopathy is none of these. Several opined something along the lines of "alternative medicine" and a few knew that it had something to do with the supposed healing ability of extremely dilute solutions. And one gave a quick, succinct answer: "Homeopathy is an affront to science." Bang on!

Homeopathy has nothing to do with herbal remedies, many of which have legitimate uses. Rather it is a practice hatched in the dark ages of science based on the idea that substances that cause symptoms in a healthy person can cure those same symptoms in an ill person. Onions, which make eyes itchy and tearful, can be used to relieve the symptoms of hay fever. There is no logic to this, but this is not where it stops. Homeopaths, defying everything we know about toxicology, believe that diluting a solution containing a homeopathic remedy increases its potency. In fact, to potentiate the remedy, dilutions are carried out to an extent that the final product in most cases doesn't even contain a single molecule of the original "remedy."

Modern-day homeopaths have to admit this, and use the argument that the sequential dilutions and the tapping of the solution against a leather pillow after each dilution leaves an imprint of the original substance on the solution. This idea holds no water. Water has no memory, but even if it did, why should it remember only the substance the homeopath has added? Why is there no memory of the hundreds of thousands of other substances that water came into contact with in rivers, lakes and sewers? And why should a ghostly image of a molecule, even if such a thing did exist, have any curing ability? Here's a suggestion. Why not just add a bunch of homeopathic remedies to drinking water? They will become diluted and according to the tenets of homeopathy become more potent. Just drinking tap water should then resolve many ailments.

Obviously it's easy to make fun of homeopathy. The concept is absurd. But millions of people around the world do rely on homeopathic medications. "Can they all be wrong?" The simple answer is yes. Popular ideas are not necessarily right. After all, bloodletting went on for thousands of years, and at one time the notion that the earth was the center of the universe was quite popular. And today, many believe that the earth was created less than 6,000 years ago. Science, though, is not a popularity contest. It relies on facts, not on opinion. And the fact is that homeopathic medications contain no active ingredients. And more importantly, while hundreds of studies on homeopathy have been published, there are no repeated trials that have provided proof of efficacy. But while the tenets of homeopathy are marinated in pseudoscience, homeopaths can serve a useful function. They ask a plethora of caring questions and lend a sympathetic ear, both processes that can translate to a reduction in stress and anxiety as the ailment naturally resolves. Add a dose of placebo, and you've accounted for the success of homeopathy. But problems can arise. And Health Canada's Natural Health Products Directorate is partly to blame by having given an uncritical free ride to homeopathic preparations, even issuing specific homeopathic drug identification numbers.

It is hard to understand how this has happened, since the directorate's stated goal is "to ensure all Canadians have ready access to natural health products that are safe, effective and high quality, while respecting freedom of choice and philosophical and cultural diversity." Safety is not an issue with homeopathic products because they contain nothing. In this context, "high quality" supposedly means that the pills are produced in an environment free of contaminants. But what about efficacy? There is actually no requirement that homeopathic producers demonstrate this, which is lucky for them, because no proof of efficacy is to be had for homeopathic mercury, arsenic, "Berlin Wall" or, most alarmingly, homeopathic "vaccines" and mosquito repellants.

It seems that Health Canada has taken the view that freedom of choice is paramount as long as there is no safety issue. When it comes to a question of efficacy, regulators conveniently look the other way. In most cases, there is no issue because in general consumers who use homeopathic preparations do so for mild conditions such as colds or minor aches and pains. But using homeopathic pills to deter insects from biting is potentially dangerous. Yet that is exactly what Mozi-Q, marketed as a natural homeopathic deterrent, claims to do. Swallow a sugar pill and keep mosquitoes away! Not only that, it is also supposed to reduce the itching if you do get bitten.

What evidence is provided? There's talk of how mosquitoes avoid delphinium flowers, which may or may not be true. But what does that have to do with swallowing pills sprayed with an extremely dilute extract of the plant? Are the nonexistent delphinium molecules exuded through the skin? And itching is supposedly relieved because a pill contains a trace of stinging nettle extract? According to the perverse theory of homeopathy, nettle causes stinging on contact with skin and therefore when diluted is a simple remedy for the same sensation. Simply asinine.

Mozi-Q also cites a reference to some homeopath who carried out a study in the 1960s, a study that cannot be found. What else? Supposedly Mozi-Q was tested over four seasons on people. How? Was there a control group? Why weren't the results published? Pity the poor person who goes into a mosquito-infected area thinking he or she is protected from bites by having swallowed a sugar pill. And remember that mosquito bites can be more than minor annoyances: they can transfer disease, such as that caused by the West Nile virus. And Health Canada thinks this is all right. It isn't.

Given that the theory of homeopathy is based on the idea that the more dilute a preparation, the more potent it is, will you overdose if you forget to take your pill? Of course not. The only risk with homeopathy is an overdose of nonsense.

PETA'S "SCIENCE" IS FOR WING NUTS

Gentlemen, don't look now, but if you are coming up short in your shorts, it may be because your mother ate too many chicken wings while she was pregnant. At least that was the message delivered to the organizers of the National Buffalo Wing Festival by the extremist animal rights activist group People for the Ethical Treatment of Animals. PETA claims, "Findings published by the Study for Future Families showed that eating poultry during pregnancy may lead to smaller penis size in male infants." Actually the study showed no such thing. There is no mention of poultry at all. PETA is guilty of spreading junk science.

And just what is junk science? Unfortunately that term is often bandied about these days by people who use it in a derogatory fashion to dismiss any research with which they don't happen to agree. But let me offer a scientific definition of junk science. I see it as any argument that claims to have greater support than the evidence actually justifies, usually to advance a political or commercial agenda or to buttress a personal conviction. Now let me explain why PETA's linking poultry consumption with male shortcomings is for the birds.

The study referred to by PETA investigated the effects of prenatal exposure to chemicals called phthalates on the development of the male reproductive tract. Why should this be of any interest to researchers? Two reasons. First, we all have detectible levels of phthalates in our blood and urine, which is no great surprise since these chemicals find wide application in food packaging, medical devices, automobile interiors, adhesives, gloves, textiles, toys, flooring, wall coverings, paints and personal care products. Second, some phthalates exhibit hormone-like properties, which is always a concern given that minute changes in hormonal activity can have major health effects.

Phthalates aroused the curiosity of scientists when a study by Professor Shanna Swan of the University of Rochester discovered an association between the ano-genital distance in male rats and their mothers' exposure to phthalates. The mental image of a scientist using calipers to measure the distance between the anus and the genitals of a rat may well elicit chuckles, but this distance is indeed a function of circulating male hormones. The "Study for Future Families" was sparked by the rat experiments and was designed to explore the possibility that, at least when it comes to phthalates, humans are basically giant rats.

Researchers analyzed urine samples collected from the mothers of eighty-five boys during pregnancy for nine different phthalate metabolites to give an indication of exposure. Then between the ages of two and thirty-six months, pediatricians measured the boys' ano-genital distance, the size of their testes and the width and length of their penis. An inverse association was found between the urinary concentration of five metabolites and ano-genital distance. As far as penis size goes, there was an association between width and the metabolites of diethylhexylphthalate (DEHP), one of the most common phthalates, but no association at all with length.

And how do chicken wings enter into this picture? Via a spurious scheme hatched by PETA's publicity machinery. Chickens, like virtually everything else we eat, do contain some phthalates. These may originate from packaging, butchers' plastic gloves, water pipes in henhouses or a myriad of other sources. So, here is PETA's specious argument: phthalates are associated with a smaller penis, chickens contain phthalates, therefore eating chicken wings during pregnancy leads to a smaller penis. What's wrong with this picture? For one, there is no study that has examined chicken-wing consumption and penis size! Suggesting that indulging in chicken wings leads to smaller penises is not backed by evidence and is simply junk science.

PETA's venture into junk science of course does not mean that there are no real issues with phthalates. Various studies have linked phthalates with respiratory issues, allergies, insulin resistance, sperm damage, decrease in sperm count, reduced levels of testosterone and thyroid hormones and increased waist circumference. These are mostly laboratory and animal studies the relevance of which to humans is questionable. That's just why researchers are following the health status of people in Taiwan with keen interest, hoping to learn something from the revelation in 2011 that some unscrupulous food producers had replaced palm oil with DEHP in the formulation of "clouding agents" used to impart a cloudiness to fruit drinks in order to make them resemble natural fruit juice as closely as possible. Such artificially colored and flavored beverages are of questionable nutritional value in the first place, but adulterate them with phthalates, and you may have a real issue. Especially when such products are used to make jams, jellies, popsicles, flavored teas, yogurt and dietary supplements. And why the adulteration? Money. Phthalates are cheap and allow for longer shelf life.

As one tainted food showed up after another, fear and panic spread, especially among pregnant women who had heard about the possible connection between the "plasticizer" contaminant and possible sex organ malformation in their offspring. Taiwan's international image also suffered a blow, as many of the adulterated foods were exported to other countries, including Canada and the U.S. And then of course there was the huge financial burden of a food recall. As expected, a thorough investigation was launched, generating a great deal of anger with the finding that the adulteration had probably been going on for decades.

Figuratively it was a sickening situation. Whether it turns out to be literally sickening as well remains to be seen. So far no adverse effects have been linked to the scandal, but when it comes to hormone-like chemicals the effects may be subtle and not immediately apparent.

The extensive food adulteration in Taiwan was a real issue, not like PETA's silly clucking about pregnant women being advised "to think twice before chomping on chicken wings, or their sons could come up short." The only thing that comes up short is PETA's junk science–based argument.

THERE'S A BIG HOLE IN CHEERIOS' BOAST

Make no mistake about it. General Mills's introduction of Cheerios sporting the label "Not Made With Genetically Modified Ingredients" is a mere marketing ploy and has nothing to do with health or nutrition. Let's start the dissection of this blatant attempt to capitalize on the anti-GMO paranoia by looking at the main ingredient in Cheerios: oats. Samuel Johnson, the eighteenth-century writer who compiled the first authoritative dictionary of the English language whimsically defined oats as the grain "eaten by people in Scotland, but fit only for horses in England." A clever Scot supposedly retorted, "That's why England has such good horses, and Scotland has such fine men!"

Modern science, as it turns out, supports the ancient Scotch penchant for oats. A form of soluble fiber in the grain known as beta-glucan has been shown to reduce levels of cholesterol in the blood, which in turn is expected to reduce the risk of heart disease. You couldn't tell this by the Scottish experience, though. Scotland has one of the highest rates of heart disease in the world. It seems all that haggis, refined carbs and a lack of veggies is too great a challenge for Scottish oats to cope with. Actually you need at least 3 grams of beta-glucan daily to have any effect on blood cholesterol and that translates to roughly a cup of cooked oat bran or a cup and a half of oatmeal. Or about three servings of Cheerios. And that makes the cholesterol-lowering claims prominently featured on the Cheerios box ring pretty hollow. There are far better ways to reduce cholesterol than gorging on Cheerios.

At least, though, the cholesterol-lowering claim has some scientific merit. The "no GMO" claim has none. To start with, there are no genetically modified oats grown anywhere, at least not in the current sense of the term, which refers to the splicing of specific foreign genes into the DNA of a seed. Such recombinant DNA technology is generally used to confer resistance to herbicides or protection from insects, but resistance to drought and enhancement with nutrients hold great potential. Although it is this new-fangled technology that garners attention these days, the fact is that virtually everything we eat has been genetically modified in some fashion over the years, either by traditional crossbreeding or through the use of chemicals or radiation, both of which can scramble the genetic material in crops. The latter processes are based on the hope that a useful mutation will occur by chance, but basically it comes down to a roll of the dice. Just do enough experiments and a valuable mutant may surface. Radiation breeding has produced many varieties of rice, wheat, peanuts and bananas that are now widely grown. If you are eating red grapefruit, or sipping premium Scotch whisky made from barley, you are enjoying the products of this technology.

So if "genetically modified" oats do not exist, what sort of monsters is General Mills protecting us from? As is the case with any commercial cereal, Cheerios contains a number of ingredients with nutritious whole grain oats at the top of the list. Next come modified cornstarch and sugar. It is to these two ingredients that General Mills refers when it talks about "GMO-free." Much of the corn and some of the sugar beets grown in North America are genetically modified to resist herbicides and ward off insects. But by the time the highly processed starch and sugar extracted from these plants reach the food supply, they retain no vestige of any genetic modification. There is no way to distinguish the starch or sugar derived from genetically modified plants from the conventional varieties. The GMO-free Cheerios will not differ in any way from the currently marketed version except that the price may eventually reflect the greater cost of sourcing ingredients from plants that do not benefit from recombinant DNA technology.

The reason for the addition of sugar to Cheerios, actually in small doses compared with other cereals, is obvious. But why is cornstarch added, and why is it modified? Nobody likes soggy cereal, and a thin layer of modified starch sprayed onto the little "O"s helps keep the interior dry. The modification in this case has nothing to do with genetic modification. Starch is a mixture of essentially two polymers, or giant molecules, both composed of units of glucose joined together. In amylose, the glucose units form a straight chain, while in amylopectin, the main glucose strand features many branches of shorter glucose chains. The properties of any starch depend on the relative proportion of amylose and amylopectin as well as on the degree of branching.

Starch has many uses in the food industry. It can thicken sauces, prevent French dressing from separating, substitute for fat or keep cereals dry. But these uses require starches of specific composition, either in terms of the length of the glucose chains or the degree of branching. In other words, the native starch has to be "modified" by treatment with acids, enzymes or oxidizing agents. There is no safety issue here — modified starches are approved food additives. Of course that doesn't prevent scientifically illiterate alarmists from scaring the public by blathering on about modified starch being used as wallpaper glue and insinuating that any food made with it will literally stick to our ribs. The modified starch used in glue, namely a carboxymethylated version, is not the same as used in food, but even if it were, so what? Just because water can be used to clean garage floors and is found in tumors doesn't mean we can't drink it.

Talking about washing garage floors, Cheerios also contains tripotassium phosphate, a powerful cleaning agent. It is added in small amounts to adjust the acidity of the mix used to formulate the cereal. This too has raised the ire of some ill-informed activists who do not realize that we consume all sorts of naturally occurring phosphates regularly in our diet. Quacking about the dangers of tripotassium phosphate in Cheerios makes about as much sense as hyping Cheerios that are "Not Made With Genetically Modified Ingredients."

I'LL PASS ON AUTOURINE THERAPY

I've often expressed skepticism about the plethora of beverages being promoted these days that claim to energize, calm, heal or detox our chemically ravaged bodies. "Don't knock it till you've tried it," I'm often told. So I've gamely downed glasses of noni juice, goji juice, acai juice, vitamin water, oxygenated water, angel tea and various homemade concoctions including the Master Cleanse, made by mixing lemon juice, Cayenne pepper and maple syrup as recommended by noted nutritional expert Beyoncé. But I've drawn the line at giving "liquid gold" a shot. When I pee into a cup, it is for sending a sample to a lab to be analyzed for creatinine, blood, proteins, ketones and glucose, all of which can indicate a problem if present in abnormal amounts. But as far as "autourine therapy" goes, I opt to pass.

Mercifully I was oblivious to this bizarre, albeit intriguing practice until 1989, when I received a letter from a woman about a therapy that "heals all human beings' illnesses." She had become persuaded about the effectiveness of this "elixir of life" after reading a document, a copy of which she enclosed for my perusal. Would I please help her, she implored, "to save millions of lives with this product that everybody possesses naturally and which God gave us for a medical purpose?" Well, with millions of lives at stake, I figured I better at least have a look at the "data" I was sent. It didn't take long before I got pissed off.

The introduction went like this: "The human race would benefit immeasurably if the medical profession was ended. The proof will be found by observing in big cities and towns, where with increase in the number of doctors, there has always been an enormous increase in the number of patients suffering from various diseases including cancer, heart disease, tuberculosis and diabetes." The document then proceeds to propose a solution to the misery caused by doctors. Sorrows can be drowned in a daily swig of urine! Not only does this remove waste products and toxins, it stimulates the body's defensive mechanism.

So why has this magic potion not been more widely publicized? Could it be that, as urophagists (that's the technical term for urine drinkers) suggest, doctors and Big Pharma have conspired to keep the life-saving information under wraps because "with urine there is no more need for medication or surgery since it kills illnesses in such a short time that doctors are afraid they will lose their jobs." Well, if this were so, doctors should have been weeded out long ago because people in India and China have been imbibing from the Golden Fountain for at least 5,000 years! And lest you think the practice is limited to uneducated members of the population, Indian Prime Minister Morarji Bhai Desai claimed in a 1978 interview on the American news program 60 *Minutes* that drinking urine was the perfect medical solution for the millions of Indians who cannot afford medical treatment. He went on to attribute his own good health to indulging regularly. Apparently he wasn't harmed by the practice, as he lived to the ripe old age of ninety-nine.

Perhaps surprisingly, urine is usually quite safe to consume. Bacteria may be present in the urethra, but unless there is an infection, these are generally washed out in the first few seconds of urination, which is why urine samples for analysis should be taken from mid-stream. As far as recycling urine in a situation when no drinking water is available, well, that's not a good idea. Like seawater, urine has a high mineral content and actually can cause further dehydration.

Dehydration is not an issue when urine is consumed for its supposed medical benefits. But can there really be benefits? Attendees at the World Conference on Urine Therapy think so.

There have been five such events to date, each one attracting an audience of hundreds, who, with appropriate autourine pee breaks, listen to physicians and scientists give evidence about their clinical work with patients as they "aim to help suffering people understand that urine is not a toxic waste but a wide spectrum healing agent, not matched by any other medication."

Although the speakers' interpretation of what constitutes evidence is rather imaginative, they do address a broad spectrum of topics. At the various meetings, there have been discussions of dosage with some proponents suggesting that four-day-old pee is more potent, others claiming that a drop placed under the tongue with an eye dropper is just as effective, conjuring up some confused analogy to homeopathy. One speaker claimed that urine should be "ionized," and he described a homemade contraption powered by a solar panel to impart the therapeutic properties. Another reported on the use of camel urine for some abdominal problems, and there was talk of pigs reaching market size faster if reared on their own fermented urine. There have been anecdotes galore about people being helped in every imaginable condition, including AIDS, allergies, asthma, flu, snake bite and menopausal symptoms, which apparently are best treated with subcutaneous injection of 1 mL of urine once a week for four to six weeks. How is all this supposed to work? We're informed that traces of substances that cause illness are secreted in the urine, and when reintroduced into the body they trigger the production of antibodies that fight disease. They also trigger skepticism.

I was particularly intrigued by a report about "Plant Urine," never having considered that plants actually voided, although I can affirm that they have been voided upon. Turns out that "plant urine" is the water that "the plant with its root system filters and lifts from great depths." I was gratified to learn that "no external water sources or artificially processed water is used, ensuring the water contains no unfavorable memories of artificial processing." Sure wouldn't want to consume psychologically disturbed plant urine. Still, I think I would choose it over "The Nectar of Life."

I wonder if eyebrows would be raised if I attended the next World Conference on Urine Therapy in order to satisfy my thirst — only for knowledge of course — and then submitted an expense report. Maybe some administrator would say "urine trouble."

SCIENCE BY PETITION

I'm sure Lisa Leake is a well-meaning young lady and a fine mother. But I think she could use a lesson in chemistry. Lisa and fellow food blogger Vani Hari are the movers and shakers behind a petition to "remove all dangerous artificial food dyes" from Kraft's classic macaroni and cheese and replace them with the natural dyes used in the United Kingdom. The argument is that Yellow #5, also known as tartrazine, and Yellow #6, known as Sunset Yellow, may be linked to health problems, whereas the natural colors, namely beta-carotene and paprika, have no such dark clouds hanging over their heads.

Food dyes have always been one of the most controversial classes of food additives because they serve only a cosmetic purpose. They do not contribute anything nutritionally and in fact may make foods of poor nutritional quality more appealing. Before getting back to Lisa and Vani's petition and allegations of danger, a little history seems to be in order.

As early as the first century, Pliny the Elder noted that wine was sometimes artificially colored, possibly with squid ink. Saffron, paprika, turmeric, beet extract and various flower petals have long been used to color foods. A peasant in the middle ages may well have eaten bread that was adulterated with lime to mimic refined flour, preferred by the rich but unavailable to the poor. King Edward I (1239–1307) took such adulteration very seriously and issued an edict that a baker guilty of such an offense should be dragged "through the great streets where there may be the most people assembled . . . with the faulty loaf hanging around his neck." Should he repeat the offense, he would be pilloried for an hour, and if he still didn't learn his lesson, "the oven shall be pulled down, and the baker made to foreswear the trade in the city for ever."

In the fourteenth century the coloring of butter was made illegal in France, and a law passed in 1574 forbade the coloring of pastries to simulate the presence of eggs. Copper compounds were commonly used to "green up" vegetables. In 1820, English chemist Friedrich Accum recounted the misadventures of a "young lady who amused herself while her hair was dressing with eating sapphire pickles impregnated with copper." The episode did not have a happy outcome. "She soon complained of pain in the stomach. In nine days after eating the pickle, death relieved her of her suffering." Accum also documented the use of red and white lead, vermilion (a mercury compound) and copper arsenite in candies designed to appeal to children. He actually published the names of the guilty manufacturers, making some powerful enemies in the process, yet adulteration continued unabated.

The situation was certainly no better in America. Pickles were bathed in copper sulfate and milk was tinged yellow with lead chromate. Indeed this was such a common process that when white milk was available, people refused to drink it, thinking it had been adulterated. But even back then there were consumer advocates. At the 1904 St. Louis Exposition, they displayed silks colored with dyes used by food manufacturers, implying that chemicals that could be used to dye fabrics were not suitable for consumption. This is what I would refer to as a "fallacy by association." Palm oil, for example, is used to make napalm, but that has nothing to do with its safety as a food. Similarly, food dyes cannot be declared dangerous just because they are made from petroleum, a substance no one would ever want to consume.

While I have no problem urging a reduction in the use of food dyes, I have a problem with unscientific arguments used toward this end, such as Lisa Leake's claim that "food companies [are] feeding us petroleum disguised as brightly colored food dyes." To dig herself an even deeper hole, she goes on to list seven reasons why she hates food dyes, with number one being that "they are made in a lab with chemicals derived from petroleum, a crude oil product, which also happens to be used in gasoline, diesel fuel, asphalt, and tar." This is senseless fear mongering from a young mom lacking any scientific background.

The allegation that we are being fed petroleum disguised as food dye is blatantly absurd. Food dyes, while synthesized from compounds found in petroleum, are dramatically different in molecular structure from any petroleum component. Furthermore, the safety of a chemical does not depend on its ancestry, but on its molecular structure. And the way to evaluate safety is through proper laboratory and animal studies with continued monitoring of human epidemiology.

Over the years, as testing methods became more and more sophisticated, and regulations more stringent, many of the dyes used historically by the food industry were removed from the market. The ones that remained, such as tartrazine and Sunset Yellow, have passed the scrutiny of regulatory agencies and are allowed in a wide variety of foods.

The legal use of a food dye, however, cannot guarantee that no adverse effect will be noted. There is always the possibility that a small subset of the population will experience some adverse reaction. Allergic reactions as well as behavioral problems in children have been noted with some food dyes, although there is a divergence of opinion about the seriousness of the problem. It is such uncertainty that has prompted the petition against Kraft. While the goal is reasonable, the suggestion that food dyes are a problem because they are man-made chemicals derived from petroleum is not.

Though it is true the dyes used by Kraft in North America

have all passed through the regulatory hoops and hurdles, there have been enough questions raised about them to give us reason to evoke the precautionary principle, which states that even if there is no proof of harm, a chemical with some potential for harm should be replaced if a safer alternative is available. Paprika and beta-carotene are indeed better choices. Of course that still begs the question of why macaroni and cheese should be colored in the first place. Anyway, all this has put me in a mood for some mac and cheese. Made from scratch. No color needed. Yum!

TIME TO STOP BABYING THE FOOD BABE

Did you know that the calcium pill you may be popping contains the same chemical found in gravestones? Or that your tasty bite of bread contains gypsum, better known as plaster of Paris? How about some toilet bowl cleaner in your cake mix? What are these nasty food and drug companies trying to do? Poison us? What an outrage! Right? Wrong!

Calcium carbonate is an effective, safe calcium supplement, and calcium sulfate is a tried and true yeast nutrient. Sodium hydrogen sulfate in combination with baking soda generates the carbon dioxide that makes cakes rise. The fact that it can also dissolve deposits in toilet bowls is irrelevant. We don't avoid pasta because flour dust can cause grain elevators to explode, or soda water because liquefied carbon dioxide is used to take stains out of fabrics. And if there is any concern about the flour additive azodicarbonamide, it should not be because it is also used in the manufacture of yoga mats or because its name may be difficult to pronounce!

That brings us to Vani Hari, of "remove the food dyes from mac and cheese" fame, an attractive young woman, who under the moniker of "The Food Babe," aims to blow the whistle on brands of foods and beverages that in her very words "are trying to slowly poison us with cheap and harmful ingredients." Hari does not have any sort of degree in food science or chemistry, but that does not seem to be an impediment when it comes to telling us that "we are getting conned by cheap, toxic chocolate" or that our beer is chock full of "shocking ingredients" or that "butter is secretly ruining our health." No, it isn't the fat or the cholesterol in the butter. It's the genetically modified organisms (GMOs) in the corn or soy that the cows are fed.

It doesn't take more than a quick perusal of the Food Babe's blog to reveal that she has no understanding of what genetic modification is all about. Just what GMOs does she think are present in butter that pose a risk? And what's with the "toxic chocolate"? Here her target is soy lecithin, used as an emulsifier. It may come from genetically modified soybeans, and as Hari exclaims, "We do know that the consumption of GMO foods poses a serious threat to our health." Actually, we know no such thing. And she says we better watch out for isinglass in beer. Why? Because it is produced from the swim bladders of fish. So what? It is just purified protein that is used to remove haziness from beer.

Now to the issue of the "yoga mat" chemical, azodicarbonamide. Yes, it can serve as a source of nitrogen, the gas that creates the tiny pockets that characterize plastic foams. It can also be used as an additive to flour where it acts as an oxidizing agent that allows protein molecules to link together to form the elastic network we know as gluten. This traps the carbon dioxide gas released by the action of yeast and helps give bread a desirable texture. At the same time, azodicarbonamide oxidizes some of the natural dark pigments in flour giving it a whiter appearance and increased consumer appeal. As with any food additive, regulatory agency approval is needed before use.

In Canada and the U.S., the chemical can be used in dough up to 45 parts per million, but it is not approved in Europe. There the issue is the small amounts of urethane and semicarbazide released when azodicarbonamide is heated. These chemicals can cause cancer when fed to animals, but only in doses far higher than any that can be found in baked goods. It is also noteworthy that urethane occurs naturally in wine, whisky and soy sauce at higher concentrations than can possibly occur as a result of azodicarbonamide breakdown in bread. Health

Canada maintains that humans are unlikely to be at risk because no experiment has shown that these breakdown products are capable of damaging DNA.

I don't know whether the trace amounts of semicarbazide or urethane pose any risk, and nobody else does either, no matter how loudly they scream. I very much doubt it, given the thousands of naturally occurring "toxins" in food to which we are regularly exposed. Piperidine in black pepper, tannins in tea, safrole in nutmeg, formaldehyde in apples, acrylamide in bread, hydrazines in mushrooms, aflatoxins in molds and alcohol itself are all classified as carcinogens, to say nothing of the heterocyclic aromatic amines and polycyclic hydrocarbons that form upon cooking. What I do know is that any decision to remove azodicarbonamide from the food supply should not be based on it being a "yoga mat chemical."

But the yoga mat argument was used very successfully by the Food Babe in initiating a petition to get Subway to eliminate azodicarbonamide from its rolls. I don't object to the chemical being eliminated — after all, it is not essential. European bakers manage quite nicely without it. Indeed, if you use high-quality flour, you don't need a dough conditioner. What I do object to, is scare tactics based on meaningless associations and "science by petition" instead of science by evidence, especially when such petitions are originated by bloggers who may have the best of intentions but lack scientific know-how. What's going to happen if the Food Babe finds out that vegetable oil can suffocate lice and vinegar can be used to kill weeds? She'll want to ban salad dressings!

While the azodicarbonamide issue is a tempest in a teapot, it does propagate the notion that we are at the mercy of incompetent regulatory agencies that fail to protect us from the flood of toxic chemicals unleashed by corporations that care only about profits and not one iota about consumer safety. Admittedly, there may be some questions raised about the risk-benefit analysis for some specific chemicals that can turn up in food, but the blanket condemnation of "chemicals" is senseless and the fears raised by many self-styled activists are unjustified.

But it is precisely such fear mongering that gets people like the Food Babe on *The Dr. Oz Show* as an expert. And isn't it curious that when *USA Today* ran a story about Pepsi planning to replace high fructose corn syrup with sugar in some beverages, they didn't approach Dr. Walter Willett of the Harvard School of Public Health or Marion Nestle of New York University for a comment? They looked to the Food Babe for wisdom.

Yes, advocacy for improved nutrition is needed. We consume far too much sugar, too few fruits and vegetables, and trans fats still lurk in some processed foods. Nutritional guidance, however, should be coming from respected authorities who base their information on the peer-reviewed literature instead of blindly parroting the unsubstantiated claims of pseudo-experts like Mike Adams of Natural News. It should also be noted that any comment on the Food Babe's blog that challenges her views is immediately deleted and results in the correspondent being banned. Not exactly in line with the proper pursuit of science. When you have a plumbing problem, you call a plumber. When you have an electrical problem, you call an electrician. Why then, when it comes to a food-related issue, which is inherently more complex, would one turn to the Food Babe?

CONCLUSION

At the beginning of this book I suggested that the development of the understanding of molecular structure was of fundamental importance in the evolution of chemistry. Let me end with a couple of examples that highlight how knowledge of molecules can lead to separating myth from fact.

Walk into a health food store and it's a good bet you can find a bottle of grapefruit seed extract. The label may declare it to be hypoallergenic, it may tell you that it is made from organically grown grapefruit, that it is vegetarian or vegan, bioflavonoid rich, "super strength" or that it has a high proanthocyanidin content. Or it may just say that it is a liquid extract or a dietary supplement. What it won't say, except in rare cases, is what it is to be used for. One supplement does identify the grapefruit seed extract as an antibiotic and recommends it be used for "candida, detrimental bacteria and microorganisms." The reason such labels are rare is because they are illegal. There is no evidence that grapefruit seed extract is of use in such infections, and if it were, the extract would have to be marketed as a drug.

Although most labels do not make such outlandish declarations, there are plenty of books, newsletters and websites that promote grapefruit seed extract as a natural antimicrobial agent to be used for sore throats or as an ear, nasal or vaginal rinse. Some marketers suggest it be taken regularly orally without saying why. In addition to the dietary supplement aisle of the health food store, you can also find grapefruit seed extract as an ingredient in cosmetics where it supposedly acts as a natural preservative. The crux of the matter is that there is no evidence that authentic grapefruit seed extract has any value at all in terms of fighting infection or acting as a preservative. There is evidence, however, that some of these products do have antimicrobial properties, but that is because they contain undeclared synthetic preservatives.

Cosmetics that are water based need preservatives. It is as simple as that. Creams and lotions make for a very hospitable environment for bacteria and fungi. Apply a cream to the face, reach back into the jar and you've contaminated it, unless a preservative is present. So manufacturers have a problem: they must use preservatives, but preservatives are also embroiled in controversy. Some, like parabens, have been linked, albeit unjustifiably, to breast cancer; others release formaldehyde, a known skin irritant and possible carcinogen. "Natural" substances are perceived to be safer, and promoting a product as containing "no synthetic preservatives" translates into increased sales. "Preserved with grapefruit seed extract" sounds better than "preserved with synthetic parabens."

But the trouble is that an extract of grapefruit seeds does not have a significant antimicrobial effect. As numerous analyses have shown, all such extracts that do have antimicrobial properties have them because they contain synthetic preservatives like triclosan, parabens, benzethonium chloride or benzalkonium chloride. Some manufacturers argue that the antimicrobial compounds detected are formed by their proprietary process of treating naturally occurring polyphenols in grapefruit seed extract with ammonium chloride. This makes no sense in terms of the chemistry and molecular structures involved. While there is no risk with the synthetics that are added to the natural extract — in fact they are responsible for the preservative effect — the labeling of grapefruit seed extract as a natural preservative is false.

And talking about false, let's discuss the allegations made about another preservative, this time one in Japanese noodles. At the turn of the millennium, a Japanese poll asked about the best Japanese invention of the previous century. "Instant noodles" was the answer. Both Japan and China have a long history of eating noodles, mostly made of wheat, although rice noodles are also popular. And then in 1958, along came Momofuku Ando with an idea. If noodles were hot-air-dried or quickly fried after they were steamed, they would last a long time and could be readily cooked by dumping them into boiling water. The instant noodles could be mixed with various flavor additives to yield a quick soup. Ramen noodles, using the Japanese term, are high in salt and some can contain a significant amount of fat. But the noodles are not "deadly."

Why should that idea even come up? Because of articles floating around the web about "what happens in your stomach when you consume packaged Ramen noodles with a deadly preservative." This bit of nonsense refers to a video that has been making the rounds about an experiment carried out by gastroenterologist Dr. Braden Kuo at the Massachusetts General Hospital. Kuo had a subject swallow a pill camera capable of transmitting images from inside the gut. He found that processed noodles churned around in the stomach longer than fresh noodles before breaking down.

This doesn't have much significance, since nutrient absorption takes place in the small intestine after food has been broken down in the stomach, but how long that breakdown takes is not important. Why the processed noodles take longer to disintegrate in the stomach can have many reasons. The moisture and fat content of the noodles can be quite different, and the gluten content, which depends on the kind of flour used, as well as the amount of kneading, make a difference, as does the shape of the noodles. But one thing that will not have an effect is the trace amount of a preservative known as tertiary butyl hydroquinone, or TBHQ, that may be present in some instant noodles.

Yet this is the ingredient that has generated all the nonsensical information being spread around the web. The claim being that TBHQ is the preservative responsible for preventing the noodles from being broken down as quickly as fresh noodles, and this represents some kind of danger to health. Both of these claims are absurd. The preservative, which in fact is not commonly used in noodles, prevents fat from going rancid, which is a process that can indeed produce toxins. The amount of TBHQ used is trivial, 0.02 percent by weight of the fat content of the food. That translates to a few milligrams, a tiny fraction of the amount that can cause any harm in an animal.

Of course the scary emails do not take amounts into account. Rather they blather on about nausea, diarrhea and ringing in the ears, which may happen at huge doses of TBHQ that cannot be attained from food. And most assuredly, TBHQ has nothing to do with the rate at which noodles decompose in the stomach. This is not an argument for eating processed ramen noodles, which are not great health-wise, particularly because of the salt content. But it is a plea for rational thinking, and the investigation of claims made by bloggers who do not know what they are talking about. Kuo himself was not troubled by his findings and says that he eats processed noodles himself. He knows all about molecules. And the myths that surround them.

INDEX